Studies in Computational Intelligence

Volume 825

Series editor

Janusz Kacprzyk, Polish Academy of Sciences, Warsaw, Poland
e-mail: kacprzyk@ibspan.waw.pl

The series "Studies in Computational Intelligence" (SCI) publishes new developments and advances in the various areas of computational intelligence—quickly and with a high quality. The intent is to cover the theory, applications, and design methods of computational intelligence, as embedded in the fields of engineering, computer science, physics and life sciences, as well as the methodologies behind them. The series contains monographs, lecture notes and edited volumes in computational intelligence spanning the areas of neural networks, connectionist systems, genetic algorithms, evolutionary computation, artificial intelligence, cellular automata, self-organizing systems, soft computing, fuzzy systems, and hybrid intelligent systems. Of particular value to both the contributors and the readership are the short publication timeframe and the world-wide distribution, which enable both wide and rapid dissemination of research output.

The books of this series are submitted to indexing to Web of Science, EI-Compendex, DBLP, SCOPUS, Google Scholar and Springerlink.

More information about this series at http://www.springer.com/series/7092

Diego Oliva · Mohamed Abd Elaziz ·
Salvador Hinojosa

Metaheuristic Algorithms for Image Segmentation: Theory and Applications

Diego Oliva
Departamento de Electrónica, CUCEI
Universidad de Guadalajara
Guadalajara, Jalisco, Mexico

Mohamed Abd Elaziz
Faculty of Science
Zagazig University
Zagazig, Egypt

Salvador Hinojosa
Facultad de Informática
Universidad Complutense de Madrid
Madrid, Spain

ISSN 1860-949X   ISSN 1860-9503 (electronic)
Studies in Computational Intelligence
ISBN 978-3-030-12933-0   ISBN 978-3-030-12931-6 (eBook)
https://doi.org/10.1007/978-3-030-12931-6

Library of Congress Control Number: 2019930979

This Springer imprint is published by the registered company Springer Nature Switzerland AG
The registered company address is: Gewerbestrasse 11, 6330 Cham, Switzerland

Foreword

In this book, the most important methods used for image segmentation are collected, especially for thresholding. The idea of the authors is to gather information in a document and provide the reader with the tools to implement new algorithms. They combine two important fields in computer sciences: artificial intelligence and image processing. It is well known that image segmentation is one of the most important tasks in computer vision systems. However, nowadays it is hard to find a compendium of information that presents the most relevant methods used by the scientific community. However, this book addresses the issues of image segmentation using metaheuristic algorithms that are part of the artificial intelligence field. The idea of combining the techniques is to improve the performance of the classical algorithms in image processing.

This book guides the reader along different and interesting implementations, but it also includes the theoretical support that permits to understand all the ideas presented in the chapter. Moreover, each chapter that presents applications includes comparisons and updated references that support the results obtained by the proposed approaches. At the same time, every chapter provides the reader with a practical guide to go to the reference sources. Meanwhile, the introductory chapters are easy to understand due to the images and the explanation of the equation and steps of the processes.

This book was designed for graduate and postgraduate education, where students can find support for reinforcing or as the basis for their consolidation; researchers can deepen their knowledge. Also, professors can find support for the teaching process in areas involving machine vision or as examples related to main techniques addressed. Additionally, professionals who want to learn and explore the advances on concepts and implementation of optimization and learning-based algorithms applied to image processing can find in this book an excellent guide for such purpose.

This interesting book has fifteen chapters that are organized considering an introduction to optimization, metaheuristics, and image processing. Here is also included a survey about the most recent studies related to the use of metaheuristic algorithms for image thresholding. In this sense, Chap. 1 provides a general overview of this book. Chapter 2 presents the concept of mathematical optimization; meanwhile, in Chap. 3, the metaheuristic algorithms are explained. Chapter 4 explains the reader the necessary concepts of image processing, and Chap. 5 introduces the reader to the image segmentation. In Chap. 6, the current trends of image thresholding using metaheuristics are described. From Chaps. 7 to 11, the reader could find the most relevant methods for image segmentation using statistical metrics as the intra-class variance, proposed by Otsu or Kapur's entropy, and the fuzzy entropy is also described.

The remaining part of this book contains the last four chapters including unconventional methodologies for image segmentation. In Chap. 12, for example, a mixture of Gaussian functions to estimate the histogram is used. Moreover, the use of multi-objective optimization algorithms for image segmentation is also explained in Chap. 13. Chapter 14 explores the machine learning alternatives for image segmentation; here, the most used clustering techniques are explained. Finally, Chap. 15 shows the theory behind the energy curve that includes the contextual information of pixels.

It is important to mention that an important advantage of this structure is that each chapter could be read separately. This book is an important reference to artificial intelligence and image processing. These areas are very important and are in constant evolution. For that reason, it is hard to collect all the information in a single book. I congratulate the authors for their effort and dedication for assembling the topics addressed in this book.

Wuhan, China   Songfeng Lu
December 2018   Huazhong University of Science
                                                                and Technology

Preface

Nowadays, digital images are present in a multitude of devices. The use of cameras has increased over the last ten years, and now they are present in many aspects of our life. Images are used for leisure activities, surveillance, autonomous vehicles, medicine, communications, etc. One of the main reasons for using image processing applications is that there is no necessity to analyze all the scenes by a human expert. For example, in industrial applications the cameras acquire images to detect defects in the products. An automatic system is programmed to automatically perform the analysis of the images. Such kind of systems has different steps, and to be completely automatic they require artificial intelligence methods. Some of these methods are applied to image processing algorithms. The idea behind image processing is to employ different methods for extracting features that allow identifying the objects in the scene. The methods used include operators to analyze the pixels in diverse ways. However, most of this kind of operations is computationally expensive.

One branch of the field of artificial intelligence includes optimization algorithms capable of solving problems where the minimization or maximization of a model is required. Optimization approaches are extensively used in different areas of engineering. They are used to explore complex search spaces and obtain the most appropriate solutions to a given problem represented as an objective function. This book focuses on lightweight segmentation methods based on thresholding techniques using metaheuristic algorithm (MA) to perform the preprocessing of digital images. The aim is providing the reader with the most representative tools used for image segmentation while examining the theory and application of MA for the segmentation of images from diverse sources. In this sense, topics are selected based on their importance and complexity in this field—for example, the analysis of medical images and the segmentation of thermal images for security implementation.

This book aims to present a study of the use of new tendencies in image segmentation. When we started working on those topics almost five years ago, the related information was sparse. Now, we realize that the researchers were divided and closed in their fields. Another difference is the current integration of digital cameras on the current lifestyle compared to a decade ago. This book explains how to use and modify different methodologies for image segmentation using meta-heuristic algorithms.

Moreover, in our research, we discover non-conventional techniques for solving the problems of segment images. The reader could see that our goal is to show that a problem of image processing can be smoothly translated into an optimization task due to the existence of a natural link between both the image processing and the optimization fields. To achieve this goal, the first four chapters introduce the concepts of optimization and image processing. The structure of the rest of the sections first presents an introduction to the problem to be solved and explains the basic ideas and concepts about the implementations. This book was planned considering that the readers could be students, researcher expert in the fields, and practitioners that are not completely involved with the topics.

This book has been structured so that each chapter can be read independently from the others. Chapter 1 presents an overview of the entire book. Chapter 2 explains the theory related to mathematical optimization. In Chap. 3 are introduced the basic concepts of metaheuristic algorithms. Chapter 4 explains some important points in image processing. Readers that are familiar with those topics may wish to skip these chapters.

In Chap. 5, the reader could find an interesting study about the methods commonly used for image segmentation and metaheuristics. Meanwhile, in Chap. 6 is presented with a survey of metaheuristic algorithms for image thresholding.

Chapter 7 explains the basic concepts of the between-class variance (Otsu's method) for bi-level and multilevel thresholding. In this chapter, the tree–seed algorithm (TSA) is used to find the best thresholds using the between-class variance as an objective function.

Chapter 8 introduces Kapur's entropy and a hybrid metaheuristic algorithm based on the combination of the salp swarm algorithm (SSA) and the artificial bee colony (ABC).

In Chap. 9, the Tsallis entropy is used for bi-level and multilevel thresholding using the electromagnetism-like optimization (EMO) to find the best configuration of thresholds for image segmentation.

In Chap. 10, the concept of minimum cross entropy (MCET) is introduced. This method is also used for the segmentation of brain magnetic resonance image (MRI) in medicine using the crow search algorithm (CSA).

Chapter 11 introduces the fuzzy entropy approaches. Here are explained the basics about type I and type II fuzzy entropy for image thresholding. Moreover, it is also shown that such methodologies can easily figure out with metaheuristic algorithms.

Chapter 12 employs the histogram approximation using a mixture of Gaussian functions to find the best thresholds in image segmentation. This problem is also addressed using metaheuristic optimization algorithms.

In Chap. 13, it is used a multi-objective optimization algorithm to find the best solutions to image thresholding. This chapter explains how to formulate the problem of multilevel thresholding using multiple objective functions.

Chapter 14 explains the theory and concepts of data clustering. This chapter aims to implement the metaheuristic algorithms to optimize the most used clustering approaches for image segmentation.

Chapter 15 introduces a relatively new concept for image thresholding called energy curve. This method includes contextual information of pixels to generate a curve with the same feature of the histogram. Here is used the Ant Lion Optimizer (ALO) to search the best thresholds using the energy curve with Otsu and Kapur objective functions.

This book has been structured from a teaching viewpoint. Therefore, the material is essentially directed for undergraduate and postgraduate students of science, engineering, or computational mathematics. It can be appropriate for courses such as artificial intelligence, evolutionary computation, and computational intelligence. Likewise, the material can be useful for researches from the evolutionary computation and artificial intelligence communities.

Finally, it necessary to mention that this book is a small piece in the puzzles of image processing and optimization. We would like to encourage the reader to explore and expand the knowledge in order to create their implementations according to their necessities.

Guadalajara, Mexico   Diego Oliva
Zagazig, Egypt   Mohamed Abd Elaziz
Madrid, Spain   Salvador Hinojosa
December 2018

Contents

Chapter 1
Introduction

1.1 Introduction

In recent decades, the increasing availability of digital cameras has fostered the development of vision-based systems devoted to solving problems in areas as diverse as medicine, topology, agriculture or surveillance, among many others [1, 2]. Despite the different natures of the applications, most systems require image segmentation to separate the objects present in the image. As a result, image segmentation has become an attractive research topic. One of the most widely used segmentation methods, due to its independence from image size and ease of implementation, is image thresholding (TH) [3]. In general, TH approaches separate the pixels in the image according to their histogram, the value of a given pixel, and one or more intensity values, known as thresholds [4]. Originally the TH was developed for the separation of foreground and background in an image receiving the name of binarization. Subsequently, the formulation was expanded to incorporate more than two classes, leading to the definition of the multilevel threshold (MTH) [5]. Both approaches consider as a conceptual basis the frequency of occurrence of each intensity value in the image to generate a histogram; then, the threshold values divide sections of the histogram to separate different classes. Thus, the TH problem can be summarized as the search for the best threshold values that segment an image. In the case of binarization, the search for the best threshold value can be done exhaustively; that is, testing all possible thresholds until the threshold value that best separates the image is found. However, the MTH cannot be treated exhaustively because with each threshold value added to the image, the complexity of the search grows exponentially [6–8].

To avoid the necessity to try all possible combinations of threshold values, optimization techniques have been adopted in the literature that allow finding the solution to complex problems by minimizing or maximizing a quality criterion. In this way it is possible to reduce the time it would take to find an optimal solution. Within the family of optimization algorithms, it is possible to find a wide variety of approaches that aim to find the optimal solution for a given problem. Among these, it is possible

© Springer Nature Switzerland AG 2019
D. Oliva et al., *Metaheuristic Algorithms for Image Segmentation:
Theory and Applications*, Studies in Computational Intelligence 825,
https://doi.org/10.1007/978-3-030-12931-6_1

to find approaches belonging to the group of collective intelligence [9]. Metaheuristic Algorithms (MA) consider the use of agents capable of efficiently exploring some of the viable solutions to the problem being analyzed. MAs are composed of populations of solutions that move or evolve during an iterative process until a termination condition is met. MAs establish simple rules that allow the generation of intelligent collective behaviors.

The following subsections are written to provide the reader with the intuition of the two main topics discussed in the book; image thresholding and me metaheuristic algorithms. Chapter 2 provides a brief explanation of classical optimization, Chap. 3 presents the basic concepts about metaheuristics for global optimization and multiobjective optimization. The Chap. 4 presents the theory related with digital image processing, Chap. 5 explains the segmentation of digital images and Chap. 6 introduces the concepts related with image thresholding. In Chaps. 7– 11 presents different methods based on statistical metrics to compute the best thresholds using metaheuristic algorithms. Meanwhile, in Chap. 12 is introduced a technique for image segmentation using Gaussian functions. Chap. 13 presents the implementation of a multiobjective algorithm for image thresholding. In Chap. 14 are explained some clustering techniques commonly used for image segmentation and finally in Chap. 15 is presented a method for image segmentation that includes the contextual information of pixels.

1.2   Metaheuristic Algorithms for Optimization

The area of optimization comprises an enormous collection of mathematical techniques designed to find the optimal values (minimum or maximum) of analyzed problems. For this purpose, each problem needs to be written as an equation that dictates the behavior of the problematic. Classical optimization takes advantage of calculus tools such as the differentiation to identify the gradient of the function. Following the gradient up or down according to the nature of the problem it is possible to identify optimal values that minimize or maximize the function. However, gradient-based techniques cannot be applied to all problems as some condition need to be met. For example, the second derivative must exist, the function needs to be continuous, and smooth. Another issue that can harm the performance of gradient-based approaches is multimodality. In this case, classical algorithms can be trapped into suboptimal solutions. In Fig. 1.1 is easy to see how a gradient-based method could be trapped in a local optimal due to the complexity of the function. To circumnavigate those limitations, many techniques started using randomness in the process leading to the formulation of stochastic approaches. One of the first stochastic methods was the gradient-descent method with random restarts [10]. However, with only the incorporation of randomness, the search process could waste time and computer resources evaluating solutions that ultimately did not contribute to the result. To accelerate the search, many problem-specific heuristics (rules) were used, enhancing the behavior of the optimizer.

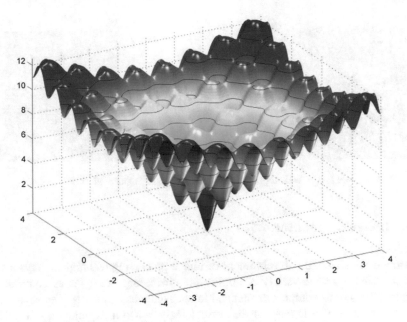

Fig. 1.1 Ackley function

The idea of Metaheuristic Algorithms (MAs) started not too long ago as an extension of stochastic optimization tools to as classic methods failed to provide good enough solutions on reasonable time for specific problems. Metaheuristic Algorithms provide rules and operators that can be applied to any problem with acceptable results. One of the first MAs reported on the literature is the Simulated Annealing algorithm (SA) [11] which takes "rules" from the process of annealing where metal is heated until a certain point, and then the temperature is dropped slowly. In SA the "temperature" determines how far the solution can move from the previous location; first, a high temperature produces large jumps resulting in a good exploration of the search space, while low temperature contributes to the refinement of the solution with only small displacements. Afterward, many other metaheuristic approaches were published where the Particle Swarm Optimizer (PSO) [12] takes an important role. PSO is a population-based metaheuristic algorithm which emulates the behavior of flocks of birds or schools of fishes. Contrary to SA, PSO explores multiple solutions (particles) at each step instead of only one. Besides, PSO models the social interactions in the swarm as a mechanism to share information between the particles to perform a better search. The PSO has a relevant role on MAs as it could be considered the basis of many algorithms that improve PSO by considering different metaphors and operators following an iterative scheme.

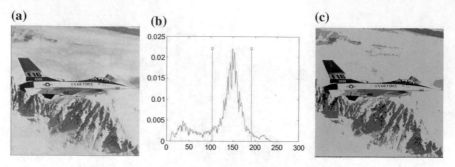

Fig. 1.2   a Original image. **b** Histogram and two thresholds. **c** Segmented image

1.3   Image Segmentation

As previously stated, the development of image processing techniques plays a significant role on a wide variety of applications including examples as common as picture editors or as complex as medical images diagnosis systems. The segmentation of images is a basic task for the correct analysis and interpretation of a scene. It very common to be used as a pre-processing stage in various vision applications, being especially useful for extracting objects of interest in an image while neglecting the rest of the scene. One of the most used techniques of image segmentation is the multilevel thresholding, which aims to group into a finite number of classes the pixels that share characteristics, in this case, their intensity. For this purpose, the following values are proposed that will act as boundaries between the segmented classes. In this way, it is possible to reduce redundant information in an image or extract structural information, which significantly facilitates and improves processing results in many tasks such as pattern detection, compression, and identification of regions of interest. Multi-level thresholding can be formulated as an optimization problem, where the values used as thresholds over the histogram of the image are searched for, while the quality of the segmented image improves. In Fig. 1.2. an image is segmented using a Multi-level thresholding method with two thresholds. The image of the segmented histogram can be observed in Fig. 1.2b).

1.4   Summary

In this chapter, a brief description of the problematics analyzed in this book is presented. The emphasis of the chapter resides on Metaheuristic Algorithms an image segmentation to provide the reader with a clear idea of the contents of the book. The disadvantage of using MAs is the need to adapt the problem to be solved from the point of view of optimization. In other words, it is necessary to define the objective function. In the case of image segmentation, it is possible to find a wide variety of implementations based on EAs with varying degrees of success, some of which are discussed in this book.

References

1. Zaitoun NM, Aqel MJ (2015) Survey on image segmentation techniques. Procedia Comput Sci 65:797–806. https://doi.org/10.1016/j.procs.2015.09.027
2. Zhang YJ (1996) A survey on evaluation methods for image segmentation. Pattern Recognit 29:1335–1346. https://doi.org/10.1016/0031-3203(95)00169-7
3. Gonzalez RC, Woods RE (1992) Digital image processing. Pearson, Prentice-Hall, New Jersey
4. Sahoo P, Soltani S, Wong AK (1988) A survey of thresholding techniques. Comput Vis Graph Image Process 41:233–260. https://doi.org/10.1016/0734-189X(88)90022-9
5. Kumar S, Kumar P, Sharma TK, Pant M (2013) Bi-level thresholding using PSO, artificial bee colony and MRLDE embedded with Otsu method. Memetic Comput 5:323–334. https://doi.org/10.1007/s12293-013-0123-5
6. Bahriye A (2013) A study on particle swarm optimization and artificial bee colony algorithms for multilevel thresholding. Appl Soft Comput 13:3066–3091. https://doi.org/10.1016/j.asoc.2012.03.072
7. Hammouche K, Diaf M, Siarry P (2010) A comparative study of various meta-heuristic techniques applied to the multilevel thresholding problem. Eng Appl Artif Intell 23:676–688. https://doi.org/10.1016/j.engappai.2009.09.011
8. Liao PS, Chen TS, Chung PC (2001) A fast algorithm for multilevel thresholding. J Inf Sci Eng 17:713–727
9. Blum C, Merkle D (2008) Swarm intelligence: introduction and applications. Springer
10. Goldfeld SM, Quandt RE, Trotter HF (1966) Maximization by quadratic hill-climbing. Econometrica 34:541. https://doi.org/10.2307/1909768
11. van Laarhoven PJM, Aarts EHL (1987) Simulated annealing. Simulated annealing: theory and applications. Springer, Netherlands, Dordrecht, pp 7–15
12. Kennedy J, Eberhart RC (1995) Particle swarm optimization. In: Proceedings of ICNN'95 - International conference on neural networks, vol 4, Perth, WA, Australia pp 1942–1948. https://doi.org/10.1109/ICNN.1995.488968

Chapter 2
Optimization

2.1 Introduction

Over the last decades, the area of optimization has contributed to finding solutions on complex problems on a wide variety of topics. Engineering applications have found optimization tools to assist the design of a wide variety of products and technologies where a specific criterion is desired to be as low or as high as possible. For example, the shape, position, orientation, and internal design of a solar cell can be modified to maximize the generated energy [1, 2]. Nowadays, the computer power available has increased, many applications can be simulated computationally, and the time and effort required to enhance designs can be significantly reduced, as prototyping can be accelerated by describing the process in a model or function suitable for minimization or maximization.

In this chapter, first, the general concepts of optimization are introduced to provide the reader with the essential information required to understand what an optimization problem is and how can be described. Then a brief discussion of classical methods for optimization is presented emphasizing the gradient-based approaches. Then, other classical approaches are discussed by enumerating the main characteristics of each method.

2.2 Optimization

The word optimization means looking for the best way to carry out an activity. From the mathematical point of view, a function is optimized when the arguments x are found that manage to generate the minimum or maximum possible result for the given function [3]. The optimization of a mathematical function is useful when this

© Springer Nature Switzerland AG 2019
D. Oliva et al., *Metaheuristic Algorithms for Image Segmentation:
Theory and Applications*, Studies in Computational Intelligence 825,
https://doi.org/10.1007/978-3-030-12931-6_2

function models a process of interest. In this case, the function can be called a cost function, objective function, or aptitude function. The optimization is generically described as:

$$x^* = \arg \min_{x \in S} f(x)$$

$$(2.1)$$

In the previous notation, S is used to group all possible solutions of the optimization problem as the solution set. The x arguments are called solution or design vector and the function f is used to measure its "quality."

However, most real-life optimization problems require more elements to properly model the process to be optimized [4]. The first requirement is the presence of a solution x with one or more design variables, where each design variable corresponds to a parameter or a solution to be found. If more than one design variable exists, then the solution is called decision vector $\mathbf{x}$. The second element of the optimization process is the objective function $f(x)$. However, the problem could have more than one objective function there are multiple goals. The third element is the solution space S which comprises the feasible solutions that could be applied to the problem modeled by the objective function. On every decision variable, at least two restrictions are required to bound the search space inside a lower and upper limit. The fourth element is the group of constraints that define the search space S. However, the constraints are not limited to bounds, as there could be inequality constraints that can help to limit the search only on feasible regions.

The function $f(\mathbf{x})$ can behave non-linearly with respect to the decision variable of $\mathbf{x}$. Due to this behavior, optimization methods work iteratively to perform an effective exploration of the search space S [5]. In this book, two classes of optimization algorithms are discussed; classical methods and metaheuristic algorithms (MAs). Classical methods mostly are based on the gradient of the function to guide the optimization process. On the MAs, the functional information of the derivative is not required. Instead, MAs implement sets of rules called heuristics to direct the search procedure. One of the main drawbacks of classical optimization methods is their dependence to the differentiation operator and their sensitivity to complex search spaces. As a result, classical methods can only work on applications with objective functions where the second derivative exists and the function is unimodal.

Optimizers only provide decision variables to the system and observe the results. The optimizers then iteratively and stochastically change the system inputs based on the feedback obtained at the time to the satisfaction of a final criterion. The process of modifying decision variables or solutions is usually defined by the corresponding mechanism of each algorithm. Depending on the relationships between inputs and outputs, the solution space can be unimodal or multimodal. In a unimodal space, there are no local solutions (hills or valleys). In the multimodal space, however, there are multiple local solutions, one of which is the global optimal. In addition, some solution spaces may have multiple global optimums with equal fitness values. An optimization algorithm should be able to discover isolated regions and identify promising optimums. It should be noted that each search space has limits, which can be considered as a set of constraints that limit the upper and lower ranges of the parameters. An optimization algorithm must be equipped with adequate mechanisms to relocate solutions that violate the allowed regions of the search space [6].

When optimizing non-linearity and multimodality are the main problems since they cause traditional methods, such as gradient ascent, to reach optimal solutions. Another problem arises when the number of decision variables grows considerably, creating spaces for huge solutions that would take a large number of computational resources for exhaustive exploration.

2.3 Gradient-Based Optimization

Classical methods for global optimization include tools taking advance of the properties of the objective function. For this purpose, theory from calculus is applied to generate iterative strategies capable of identifying the global optimum on objective functions. Among all optimization strategies, gradient-based methods are widely used due to their simplicity and effectiveness. The entire family is based on a method called gradient ascent or descent (according to the nature of the problem). The gradient descent method is the most frequently used method for non-linear optimization despite its slow convergence rate [7].

The methods encode the set of decision variables of the problem into a vector. In the start, values of the decision variables are randomly selected and grouped into the solution $\mathbf{x}^0$. After that, the decision vector is iteratively modified until a fixed number of iterations is achieved hoping to find the optimal solution g^*. The modification performed to the design vector is controlled by the expression on Eq. (2.2).

$$\mathbf{x}^{t+1} = \mathbf{x}^t + \alpha \cdot \mathbf{g}(f(\mathbf{x})) \tag{2.2}$$

where t indicates the number of the current iteration and α is the step-size of the search. The term $\mathbf{g}(f(\mathbf{x}))$ indicates the gradient of the objective function $(\mathbf{x})$. The gradient $\mathbf{g}$ of a function $f(\mathbf{x})$ at the point $\mathbf{x}$ indicated the direction on which the function $f(\mathbf{x})$ presents its greatest growth. On a minimization problem, the direction of the descent can be obtained as the contrary direction of $\mathbf{g}$ by multiplying by minus one. In this context, the condition $f(\mathbf{x}^{t+1}) < f(\mathbf{x}^t)$ is satisfied which implicates that every new solution is better than the previous.

The gradient of a multidimensional function $f(\mathbf{x})$, $(\mathbf{x} = ((x_1, \ldots, x_d)) \in \mathbb{R}^d)$ models the variation of a function with respect to one of its dimensions d. In this way, the gradient g_{x_1} expresses how the function $f(\mathbf{x})$ varies concerning x_1. Formally the gradient is defined as:

$$g_{x_1} = \frac{\partial f(\mathbf{x})}{\partial x_1} \tag{2.3}$$

Since the objective function is usually unknown in real-life applications, the analytic gradient of the function cannot be obtained. Thus, the computation of the gradient is determined by numeric methods. For example, the numerical determination of the gradient g_{x_1} is done according to the following steps:

1. A new decision vector $\tilde{\mathbf{x}}$ is generated, with a copy of the values of $\mathbf{x}$ except on the x_i component. This value will be replaced by the term $x_i + h$, where h is a small value.

$$\tilde{\mathbf{x}} = (x_1, x_2, \ldots, x_i + h, \ldots, x_d) \tag{2.4}$$

2. The gradient g_{x_1} is calculated according to:

$$g_{x_1} \approx \frac{f(\tilde{\mathbf{x}}) - f(\mathbf{x})}{h} \tag{2.5}$$

The family of gradient-based optimization techniques is one of the most used approaches to optimize problems.

2.4 Other Classical Approaches

In this section, a brief discussion of other classical approaches is presented. The considered approaches are Newton's method, line search, linear programming, Simplex, and Lagrange multipliers. The comparison is presented in Table 2.1 for readability.

2.5 Summary

On this chapter, the basic elements of the optimization theory were presented. First the essential elements of an optimization problem where discussed. Later, the elements classical methods were mentioned. The definition of the gradient descent method for minimization is explained due to its relevance to classical approaches. Finally, the importance and characteristics of other classical approaches is summarized.

Table 2.1 Comparison of classic optimization algorithms

Method	Characteristics
Newton's method	As one of the algorithms based on the gradient, Newton's method for root-finding is modified to find optimal solutions since the optimization of a given function can be translated as finding the root of the derivative of the objective function. This method is especially effective on quadratic functions since it can find the optimum in only one step. It should also be noticed that it requires the computation of the Hessian matrix (second derivative) which can be time-consuming
Line search	This is another version based on the gradient descent algorithm. In this method, the step size of the jump (α) is determined as a good enough approximation instead of trying to determine the best step size
Linear programming	Linear programming belongs to a different family of optimization methods. The main idea is to find the maximum or minimum of a linear objective under linear constraints. For this purpose, the search space is bounded by linear restrictions forming a polygon where the vertices of the polygon form the set of extreme points. It is expected that an extreme point will obtain the optimal solution to the problem
Simplex	The simplex method works by assuming that all extreme points are known. If it is not the case, then the extreme points are determined. After that, with the known extreme points it is easy to test whether or not an extreme point is optimal using algebraic relationship and the objective function. If the analyzed point is not optimal, then an adjacent point is tested
Lagrange multipliers	In this method, the basic idea is to transform a constrained method into an unconstrained problem by identifying scalars called Lagrange multipliers required to map the representation

References

1. Oliva D, Abd El Aziz M, Ella Hassanien A (2017) Parameter estimation of photovoltaic cells using an improved chaotic whale optimization algorithm. Appl Energy 200:141–154. https://doi.org/10.1016/J.APENERGY.2017.05.029
2. Díaz P, Pérez-Cisneros M, Cuevas E et al (2018) An improved crow search algorithm applied to energy problems. Energies 11:571. https://doi.org/10.3390/en11030571
3. Baldick R (2006) Applied optimization: formulation and algorithms for engineering systems. Cambridge University Press, Cambridge
4. Venkataraman P (2009) Applied optimization with MATLAB programming. Wiley, New York
5. Yang X-S, Wiley InterScience (Online service) (2010) Engineering optimization : an introduction with metaheuristic applications. Wiley
6. Mirjalili S (2019) Evolutionary algorithms and neural networks. Studies in Computational Intelligence
7. Dennis JE (1978) A brief introduction to quasi-Newton methods, pp 19–52

Chapter 3
Metaheuristic Optimization

3.1 Introduction

In the previous chapter, the ability to find optimal solutions to non-linear optimization problems of gradient-based optimizers was discussed. However, those techniques guarantee the convergence of the algorithm only if the objective function has a second derivative and the landscape of the search space is unimodal [1]. In the Fig. 3.1 two landscapes of objective functions are presented; the first is the unimodal function of the sphere and the second shows a multimodal of the landscape of the Himmelblau's function.

Metaheuristic Algorithms (MAs) operate without considering the information provided by the gradient of the objective function. This fact allows MAs to be applied over complex objective functions even if the requirements of classical methods are not satisfied. Instead, MAs require rules called heuristics that can be applied over any problem (meta-heuristics) to generate successful search patterns. Such rules are often inspired by natural or social processes. However, MAs come with a drawback; since MAs work without analytic information of the objective function, MAs are slower than gradient-based methods. MAs share a stochastic nature as they operate using random variables to provide a better exploration of the search space. Nevertheless, the randomness involved makes the analytic analysis of MAs challenging. As an alternative, the comparison and assessment of MAs are mostly performed experimentally.

The organization of this chapter includes the analysis of the operation process of most MAs as a generic procedure. Then, the problematics faced by MAs are presented followed by a discussion on the classification of the MAs. After that, the chapter focuses on single-objective and multi-objective algorithms to present historically relevant methods and their peculiarities. Next, the basic version of the Particle Swarm Optimizer (PSO) is presented as a case of study since most of the MAs used along the book are inspired by PSO followed by a brief discussion of the contents of this chapter.

© Springer Nature Switzerland AG 2019
D. Oliva et al., *Metaheuristic Algorithms for Image Segmentation:
Theory and Applications*, Studies in Computational Intelligence 825,
https://doi.org/10.1007/978-3-030-12931-6_3

Fig. 3.1 The unimodal and
multimodal landscape of the
a sphere function and
b Himmelblau's function

(a)

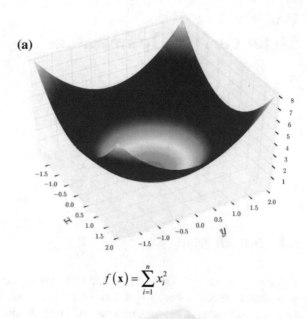

$$f(\mathbf{x}) = \sum_{i=1}^{n} x_i^2$$

(b)

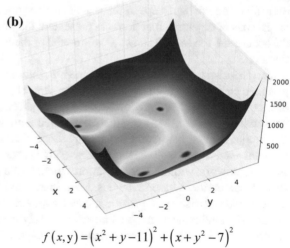

$$f(x,y) = \left(x^2 + y - 11\right)^2 + \left(x + y^2 - 7\right)^2$$

3.1.1 Generalization of a Metaheuristic Algorithm

Most MAs follow a general scheme as they are iterative process. The basic idea is to
search for an optimal solution of a given problem formulated as:

$$\min/\max f(\mathbf{x}), \mathbf{x} = (x_1, \ldots, x_d) \in \mathbb{R}^d, \mathbf{x} \in \mathbf{X} \tag{3.1}$$

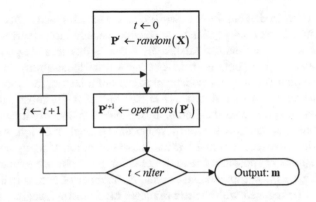

Fig. 3.2 The general process of a metaheuristic algorithm

where **x** is the design vector that encodes de decision variables of the problem, usually, a design vector is also called candidate solution. **X** is the feasible search space or solution space limited by the lower and upper bounds l_i, u_i of each one of the d design variables such as $\mathbf{X} = \{ \mathbf{x} \in \mathbb{R}^d | l_i \leq x_i \leq u_i, i = 1, \ldots, d \}$. The optimization algorithms maintain a population $\mathbf{P}^t = \{ \mathbf{x}_1^t, \mathbf{x}_2^t, \ldots, \mathbf{x}_N^t \}$ of N solutions at iteration t. On the initial state $t = 0$, all particles of the population are initialized with random positions inside **X**. On every generation, a set of operators are applied to the population $\mathbf{P}^t$ to produce a new population $\mathbf{P}^{t+1}$. The operators are the essence of the algorithm, and they control how well or bad the algorithm behaves. The quality of each solution x is evaluated using the objective function $f(\cdot)$ of the problem. It is a widespread practice that the best solution identified is stored in an external memory **m**. Figure 3.2 presents a graphical representation of the general optimization process.

3.1.2 Problematics of Metaheuristic Algorithms

In general terms, most metaheuristic algorithms have very similar frameworks. First, they begin the optimization process with the creation of an initial group of random solutions that satisfy the restriction of the problem to be solved. Then, the set containing all the solutions is iteratively evaluated by one or more target functions associated with the problem and iterate over generations to minimize or maximize the objective(s). Even though this framework is quite simple, finding solutions for real-life problems requires the consideration and addressing of several matters, from which the most important are: local optimal avoidance, the computational cost of function evaluation, constraints handling, multiplicity of objective functions, and uncertainties.

The stagnation of solutions in a local optimal is a frequent phenomenon observed on optimization algorithms, especially in real-life problems since that kind of problematics can be treated as black boxes with unknown search spaces that could contain many sub-optimal solutions [2]. Under such circumstances, the algorithm is trapped in one of the local solutions and mistakenly assumes that it is the global solution. Even if the stochastic operators of MAs improve the local ability to evade local optimal, in comparison to deterministic classical optimization methods, stagnation in suboptimal solutions can also occur in any MA. For the most part, MAs are population-based approaches that iteratively evaluate and improve a set of solutions, commonly called population or swarm, rather than a single solution. Although this also helps to improve local optimal evasion, the optimization of costly problems with MAs is sometimes not possible due to the need for many objective function evaluations. To alleviate the computational burden, the number of access to the objective function must be reduced with the use of clever mechanisms. Another difficulty of real problems is the management of the constraints imposed by the definition of the problem [3]; a simple way to model the constraints is by partitioning the search space into feasible and unfeasible regions. MAs search agents (solutions) need to be operated with adequate mechanisms to prevent the exploration of all unfeasible regions and to explore feasible areas and accelerate the finding of the global optimal solution.

In the field of optimization, there is a theorem called No Free Lunch (NFL) [4] that logically proves that an algorithm capable of successfully solving all problems does not exist. This means that the solution to a particular problem is not necessarily the best algorithm in test functions or other real-world problems. To solve a problem, it may be necessary to modify, improve or adapt an algorithm. This makes the research field very active. That is why a rigorous and accepted classification of optimization methods as such does not exist, and there is even a great debate in the literature. However, some concepts associated with optimization problems can be categorized according to the following criteria.

3.1.3   Classification

The classification of stochastic optimization algorithms is a challenging task since a rigorous and accepted classification of optimization methods as such does not exist. In the literature there is even a great debate, however, some concepts associated with optimization problems can be categorized according to the following criteria: type of search agents, restrictions, landscape of the search space and number of objectives [5]. Figure 3.3 presents a diagram that summarizes the categories.

According to the type of search space and objective function, stochastic optimization algorithms are classified into two main categories [6]: individual or population-based. In the first category, an optimization algorithm begins with a candidate solution for the problem at hand. The solution is evaluated by the objective function and improved iteratively until the satisfaction of a stop criterion. In the second category, a group of solutions is used to search for the overall optimum of the optimization problem. In Mirjalili [7] it is stated that the benefit of single-solution can be applied

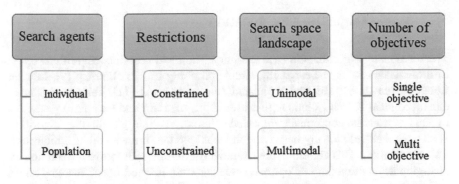

Fig. 3.3 Classification of optimization algorithms according to various criteria

when the optimization problem requires a minimum number of objective function evaluations. In single-solution algorithms, the number of iterations is equal to the number of objective function evaluations. Nevertheless, single-solution optimizers suffer from premature convergence or stagnation on local minima. The probability of an algorithm with a single solution to be trapped on some of the local solutions in real-world problems is high. On the other hand, population-based techniques are known to be highly explorational on the search space with a smaller chance of being trapped locally. If one solution is within a local suboptimal, other solutions will help to avoid it in subsequent iterations. As a disadvantage, such algorithms require more objective function evaluations and are expensive computationally.

The classification of the groups of restrictions and landscape is very straightforward. If the algorithm can handle bounds or inequality constraints it can be classified as constrained; on the contrary, if there are no limits on the search space, it will be classified as unconstrained. The topology of the search space helps to classify the algorithms that can handle unimodal or multimodal surfaces. As the name suggests it, multimodal surfaces possess multiple good solutions, from which many of them can be considered as local optima. This scenario presents additional challenges to MAs, and the operators of MAs need to be modified in order to find multiple good solutions, instead of only the global best. Most MAs can be slightly modified to register multiple goods solutions. The final criterion for classification is the number of objectives. Single-objective MAs work by minimizing or maximizing a single objective function that represents the problem at hand. The problems that need more than one criteria to be optimized are handled using multi-objective algorithms. It should be noticed that one important requirement for the use of a multi-objective algorithm the objectives must be conflictive; this means that the optimization of an objective will hurt the fitness of another objective. If the objectives are not conflicting with each other, the problem can be solved as a single objective problem by aggregating all objective functions into one. Since the topic of single-objective and multi-objective optimization is important for this book the reader will find a further explanation on the following sections.

3.2 Single-Objective Optimization

There are many algorithms in each category. The most notorious algorithms based in individuals are Tabu Search [8], simple climbing or hill climbing [9], Iterated Local Search (ILS) [10], and Simulated Annealing (SA) [11]. Tabu Search is an enhanced local search technique that uses short, medium and long-term memories to ban and truncate unpromising/repeated solutions. The simple climbing algorithm is also another individually based search technique that initiates optimization from a single solution. This algorithm then iteratively attempts to improve the solution by changing its position until an optimal value is found. Iterated Local Search (ILS) is an improved simple scaling algorithm to decrease the probability of being caught in local optimal. In this algorithm, the optimum obtained at the end of each execution is maintained and is considered as the starting point in the next execution. The SA algorithm uses a behavior that proportionally regulates to a variable called cooling factor the length of the jump that is performed in the search. Thus, when the "temperature" is high, large jumps can be made in the search space that can lead to worse solutions, while at the end of the search the temperature will be so low that it will only allow small movements that will refine the solution found. This helps SA to promote the exploration of the search space and prevents it from being trapped in the optimal premises when looking for them. In recent years, there have been many improvements in each of the above algorithms. The main objective of most of them has been to alleviate the main disadvantage of these techniques: premature convergence. Despite several improvements with successful applications [12], the nature of such techniques requires that they show less exploration and avoidance of local optimal compared to population-based methods.

Metaheuristic Algorithms work by applying an iterative process of improvement to a set of solutions until a degree condition is met. One of the most important elements required by MAs is the correct balance between operators. At some degree, all MAs perform two basic phases: the exploration or diversification and the exploitation or intensification [13]. The abrupt change of the solutions (or solution) is performed when the algorithm requires to explore the search space to avoid its stagnation on suboptimal solutions by identifying promising regions over the fitness landscape. The exploration is performed by applying large changes to the solution. However, an excess of diversification stage can prevent the convergence of the algorithm; with only the exploration phase, the optimization algorithm could behave like a random search. On the other hand, the exploitation phase helps to ensure the convergence of the algorithm. The intensification is used by applying small perturbations to the solution to refine its position trying to find small improvements of the fitness value. Similar to the diversification, if the intensification is abused the algorithm can misbehave by falling into suboptimal solutions; in this case, an algorithm with only an intensification phase can behave like a local search.

Algorithms based on populations have a specific mechanism to balance the exploration-exploitation rate. For example, the Particle Search Optimization [14] has inertial weights to maintain the tendency of particles in their previous directions and emphasizes exploration. Thus, the weights can be modified to change the balance between phases; a low inertia weight causes a low exploration and a greater tendency towards the best personal/global solutions obtained. Therefore, the particles converge towards the best points instead of navigating around the search space. In genetic algorithms, Genetic Algorithms (GA) [15], a high probability of mutation causes random changes in individuals as the main mechanism of exploration. The mechanism leading to the exploitation of the GA is the crossing operator. The crossing process causes slight random changes in the individuals and in the local search around the candidate solutions. The exploitation and exploration mechanism are different depending on the type of algorithm.

One of the major difficulties of designing and implementing MAs is providing a good balance between the exploration and exploitation phases. The simple exploratory behavior produces in most cases inferior quality solutions, as the algorithm never has the opportunity to enhance the quality of the solutions. Conversely, the exploitative behavior by itself results in retention in suboptimal solutions, as the algorithm does not apply large changes. To solve the issue of balancing the two phases many population-based optimizers adaptive operators to adjust exploration and exploitation dynamically and proportionally to the number of iterations. In other words, the algorithm is capable of performing a smooth transition from the exploration to the exploitation stage as the number of iterations increases. Another attractive technique is to boost exploration at any stage of optimization if there are no significant improvements in the best solution obtained so far [7].

3.3 Multi-objective Optimization

In many cases, real-life engineering applications cannot be model as a single-objective optimization problem. As an alternative, there are optimization methods that can handle multiple objectives simultaneously. Optimization in multi-objective search space is very different and requires special considerations compared to a single-objective optimization problem. In a problem with only one objective, there is an objective function that needs to be optimized and a single global solution must be found. However, in multi-objective problems, there is no longer a single solution to the problem, and a set of solutions must be found that represent the best relationships or compromises between multiple objectives, the Pareto optimal set.

Multi-objective evolutionary algorithms (MOEAs) are designed as optimization methods that require finding optimal solutions for two or more objectives. Unlike Evolutionary Algorithms (EAs) that are also part of MAs, MOEAs generate as results a set of solutions where each solution represents a compromise between objectives. The main idea behind multi-objective optimization was introduced by Schafer [16] with the addition of evolutionary methodologies that provide extra functionality such as local optimal evasion and gradient free search strategies. Some of the most relevant multi-objective methods include the Genetic Algorithm of Non-Dominant Classification (NGSA) [17], NGSA-II [18], Multiobjective Particle Swarm Optimization

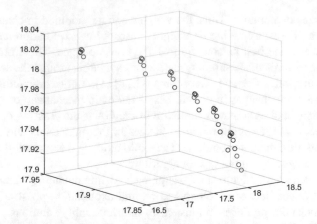

Fig. 3.4 Visualization of the solutions of the multi-objective algorithm over a three objective maximization problem, the red particles are non-dominated solutions

(MOPSO) [19], and the Multiobjective Evolutionary Algorithm based on Decomposition (MOEA/D) [20]. These methods can simultaneously optimize two or more targets while maintaining the relationship between the targets through engagement between existing targets.

The most notable difference between MOEAs and EAs is the number of objectives to be optimized; as the name suggests, the former handles two or more objectives while the latter handles only one. However, MOEAs share mechanisms and operators with EAs. Most MOEAs were designed as an extension of a successful EAs; as a result, both rely on populations of solutions to explore the search space. In each iteration, the fitness value or quality of each solution is determined by the objective function. In single target problems, the suitability value can be directly compared with a relational operator to determine which is higher or lower, while multi-target problems cannot be compared using the same operator. Given a vector function $f(\mathbf{x}) = [f_1(\mathbf{x}), \ldots, f_0(\mathbf{x})]$ and a feasible search space Ω, MOEAs are designed to find a vector $\mathbf{x} \in S$ such that optimizes the multiple objectives contained in $f(\mathbf{x})$. For this, the definitions described by Coello [19] establish a comparison framework that includes the concepts of Pareto dominance, Pareto Optimality, and Pareto Front.

Pareto dominance:

First, for minimization, the Pareto dominance indicates that the vector x dominates the vector $\mathbf{x}'$ (noted as $\mathbf{x} \prec \mathbf{x}'$) if $f_i < f_i(\mathbf{x})$ for all functions i included in the function vector f, and it exists at least one objective i such that $f_i(\mathbf{x}) < f_i(\mathbf{x}')$. In the Fig. 3.4 the solutions of a three-objective optimization problem are presented, where the black circles show solutions of the population while the red ones are non-dominated solutions.

Pareto optimality:

Then, a decision vector $\mathbf{x}^*$ is called Pareto optimal if a design vector $\mathbf{x} \in S$ such that $\mathbf{x} \prec \mathbf{x}^*$ doesn't exist. Following this idea, the Pareto-optimal set is constructed as $P^* = \{\mathbf{x} \in S\}$.

Pareto front:

Finally, the optimal Pareto front is defined as $PF^* = \{f(x) | x \in P\}$.

Thus, Zhou et al., [21] states that the ultimate goal of multi-objective optimization is to find the most accurate approach to the Pareto Optimal Front with a high degree of diversity.

The analysis of the performance of a multi-objective metaheuristic algorithm can is not as straightforward as the results generated by a single-objective algorithm. For this purpose, specific metrics such as the hypervolume designed by Zitzler were proposed [22]. It quantifies the convergence behavior of multi-objective algorithms. The principle behind it is the computation of the area, volume or hypervolume of the objective space that is dominated by the Pareto solutions obtained. The values generated by this metric are from 0 to 1 where 1 indicates the maximum hypervolume possible from the given point towards the origin of the search space. A higher value indicates better performance. Lastly, Fonseca and Flemming [23] introduced the term attainment surface, which is a boundary that separates the objective space into two regions: those objective solutions that are attained (or dominated by or equal to) the outcomes generated by the analyzed algorithms. The attainment surface graphs [24] indicate the best, the median, and the worst attainment surface. It is expected that the three of them are as close to the true Pareto as possible. In Fig. 3.5 we can observe an example of an attainment surface of a bi-objective minimization problem.

3.4 Case of Study: Particle Swarm Optimization

Complex engineering problems are not solvable by classic optimization methods on an acceptable time. Even more, the properties of the problem might not satisfy the conditions required to be solved by gradient-based methods. Those limitations have pushed the development of methodologies capable of handling complex problems. First, the incorporation of stochastic processes to optimization algorithms has provided a powerful tool that can handle the presence of local optimal and enhance the search capabilities of classical optimization algorithms. Then, the reduction of convergence times with the use of heuristics and metaheuristics has made possible solving highly complex problems in reasonable time. However, the search for better optimization algorithms did not stop there. Another milestone of the optimization community was the development of population-based algorithms also known as swarm-based intelligence.

Kennedy and Eberhart proposed in 1995 the Particle Swarm Optimization (PSO) [14]. PSO is considered the first algorithm to be classified as swarm intelligence,

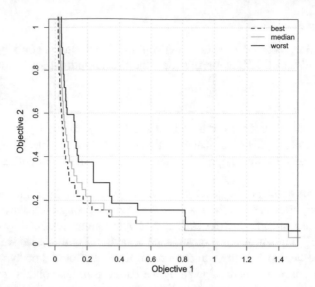

Fig. 3.5 Attainment surface of a bi-objective optimization problem

a category that comprises a large number of Metaheuristic Algorithms. The main characteristic of swarm-based metaheuristic algorithms is the use of a population of particles able to coordinate itself to perform an "intelligent" exploration of the bounded search space. Despite the fact that other approaches from the same era used populations such as Genetic Algorithms [15], the main difference resides in the interaction of the solutions in the population. On swarm-intelligence, at each generation (or iteration), the movement of each particle is not only decided by its fitness value but also by the "social" interaction with the surrounding particles or the entire swarm, while on Genetic Algorithms the only interaction between solutions is conducted by the crossover operator. The PSO algorithm was originally designed to emulate the behavior exhibited by flocks of birds and other groups of animals where each individual from the swarm acts with some degree of independence from the rest, but also considers the behavior of the population, paying special attention to the leader. On PSO, the concept of a leader is applied by selecting as leader the best solution at each generation; as the leader possess the best fitness, the rest of solutions will be drawn towards the best particle. However, to prevent premature convergence, each particle will also have some freedom to explore its location.

Computationally, the basic model of PSO initially considers a population of uniformly distributed candidate solutions, also known as particles. The quality or fitness of each particle is evaluated by an objective function given by the addressed problem, and the best particle is stored as the leader. Then, new positions for all particles are computed considering their previous position and the leader particle. For this purpose, a velocity vector is computed for each element, and then the particles are moved. After the displacement, the new positions are evaluated using the objective function and only if the new position improves the fitness of the particle over the previous location the new position is stored. Once every particle has been evaluated, the leader is updated if another particle has better fitness. The process is repeated until a stop criterion is met. Algorithm 3.1 shows the standard version of PSO while on the subsequent sections each phase is explained in detail.

Algorithm 3.1 [PSO (N, $Iter_{max}$)]

1. Input parameters
2. Initialize: set the iteration counter $t = 1$, initialize the number of N particles $\mathbf{x}^t$ uniformly in the population $\mathbf{X}$ and identify the leader as $\mathbf{g}$
3. while $t < Iter_{max}$ do
4. $\mathbf{v}^{t+1} \leftarrow \text{Velocity}(\mathbf{v}^{t+1}, \mathbf{x}^t, \mathbf{g}, \mathbf{p})$
5. $\mathbf{x}^{t+1} \leftarrow \text{Move}(\mathbf{x}^t, \mathbf{v}^{t+1})$
6. $\mathbf{x}^{t+1} \leftarrow \text{Select}(\mathbf{x}^{t+1}, \mathbf{x}^t)$
7. end while

3.4.1 Initialization

Following the general scheme of metaheuristic algorithms presented in previous sections the initialization of particles on PSO is conducted by generation random solutions following a uniform distribution bounded by the lower and upper limits of the problem to be solved. On this stage, the parameters (restrictions and gains) required by the algorithm are also set. Each particle is generated as:

$$x_k^{i,t} = l_k + rand(u_k - l_k), x_k^{i,t} \in \mathbf{x}^t \tag{3.2}$$

where $x_k^{i,t}$ is the i-th decision variable of the particle $\mathbf{x}^t$, i has as maximum value the number of particles of the population. The dimension of the problem is indexed by k, and t indicates the number of generations or iterations while l_k and u_k are the lower and upper bounds respectively for the dimensions of the search space generated by the problem. Finally, *rand* is a number randomly generated by a uniform distribution.

3.4.2 Velocity Determination

In order to obtain the new position of each particle on the search space, their velocity must be calculated. For this purpose, the previous velocity is considered; on the first generation it will be zero. Another element involved in the determination of the velocity is the position of the leader and the best particle from its vicinity. In Eq. 3.3 we can observe how the velocity is calculated:

$$\mathbf{v}^{t+1} = v^t + rand1 \times \left(\mathbf{p} - \mathbf{x}^t\right) + rand2 \times \left(\mathbf{g} - \mathbf{x}^t\right) \tag{3.3}$$

where $\mathbf{v}^{t+1}$ is the updated velocity vector while $\mathbf{v}^t$ is the previous velocity, $\mathbf{x}^t$ is he solution vector or particle, $\mathbf{p}$ contains the best position associated to the vicinity of the particle $\mathbf{x}^t$, and $\mathbf{g}$ is the global best particle or leader. On the other hand, the random numbers *rand1* and *rand2* are generated using a uniform distribution within the interval [0,1]. The two random values are incorporated into Eq. 3.3 to allow different trajectories while also considering the global and local best.

3.4.3 Particle's Movement

After the determination of the velocity, the position of the particles are updated towards their new locations on the current iteration. The movement is conducted by a simple operation where the new velocity and the previous position are combined according to Eq. 3.4.

$$\mathbf{x}^{t+1} = \mathbf{x}^t + \mathbf{v}^{t+1} \tag{3.4}$$

where $\mathbf{x}^{t+1}$ is the vector where the new position is stored, $\mathbf{x}^t$ is the previous position and $\mathbf{v}^{t+1}$ is the velocity described in the previous subsection.

3.4.4 Selection

After the movement of the particle, the new position is evaluated using the objective function to determine the new fitness. According to the result, if the fitness of the new position is worse, then the new position is discarded. The stop criterion is evaluated after this selection process is applied to every particle in the population. If the criterion is not met, the algorithm iterates again, and a new set of velocities is calculated

3.5 Summary

On this chapter, the most important theoretical topics of metaheuristic algorithms were presented. The contents of this chapter provided the necessary background to understand the implementations presented in this book. Besides, the main contribution of this chapter is emphasizing the importance of the balance between exploration and exploitation in optimization algorithms.

References

1. Bartholomew-Biggs MC (2008) Nonlinear optimization with engineering applications. Springer, Berlin
2. Addis B, Locatelli M, Schoen F (2005) Local optima smoothing for global optimization. Optim Methods Softw 20:417–437. https://doi.org/10.1080/10556780500140029
3. Coello Coello CA (2002) Theoretical and numerical constraint-handling techniques used with evolutionary algorithms: a survey of the state of the art. Comput Methods Appl Mech Eng 191:1245–1287. https://doi.org/10.1016/S0045-7825(01)00323-1
4. Wolpert DH, Macready WG (1997) No free lunch theorems for optimization. IEEE Trans Evol Comput 1:67–82. https://doi.org/10.1109/4235.585893
5. Yang X (2010) Engineering optimization. Engineering optimization. Wiley & Sons Inc, NJ, USA, pp 15–28
6. Shilane D, Martikainen J, Dudoit S, Ovaska SJ (2008) A general framework for statistical performance comparison of evolutionary computation algorithms. Inf Sci (Ny) 178:2870–2879. https://doi.org/10.1016/J.INS.2008.03.007
7. Mirjalili S (2019) Evolutionary algorithms and neural networks. Studies in Computational Intelligence
8. Glover F (1989) Tabu search—Part I. ORSA J Comput 1:190–206. https://doi.org/10.1287/ijoc.1.3.190
9. Goldfeld SM, Quandt RE, Trotter HF (1966) Maximization by quadratic hill-climbing. Econometrica 34:541. https://doi.org/10.2307/1909768
10. Lourenço HR, Martin OC, Stützle T (2003) Iterated local search. Handbook of metaheuristics. Kluwer Academic Publishers, Boston, pp 320–353
11. van Laarhoven PJM, Aarts EHL (1987) Simulated annealing. Simulated annealing: theory and applications. Springer, Netherlands, Dordrecht, pp 7–15
12. Boussaïd I, Lepagnot J, Siarry P (2013) A survey on optimization metaheuristics. Inf Sci (Ny) 237.82–117. https://doi.org/10.1016/J.INS.2013.02.041
13. Črepinšek M, Liu S-H, Mernik M (2013) Exploration and exploitation in evolutionary algorithms. ACM Comput Surv 45:1–33. https://doi.org/10.1145/2480741.2480752
14. Kennedy J, Eberhart RC (1995) Particle swarm optimization. Proceedings of IEEE international conference on neural networks, vol. 4, pp. 1942–1948. https://doi.org/10.1109/icnn.1995.488968
15. Goldberg DE, Holland JH (1988) Genetic algorithms and machine learning. Mach Learn 3:95–99. https://doi.org/10.1023/A:1022602019183
16. Srinivas N, Deb K (1995) Muiltiobjective optimization using nondominated sorting in genetic algorithms. Evol Comput 2:221–248. https://doi.org/10.1162/evco.1994.2.3.221
17. Deb K (1999) Multi-objective genetic algorithms: problem difficulties and construction of test problems. Evol Comput 7:205–230. https://doi.org/10.1162/evco.1999.7.3.205
18. Deb K, Member A, Pratap A et al (2002) A fast and elitist multiobjective genetic algorithm. IEEE Trans Evol Comput 6:182–197
19. Coello CAC, Pulido GTGT, Lechuga MSMS et al (2004) Handling multiple objectives with particle swarm optimization. IEEE Trans Evol Comput 8:256–279. https://doi.org/10.1109/TEVC.2004.826067
20. Zhang Q, Li H (2007) MOEA/D: a multiobjective evolutionary algorithm based on decomposition. IEEE Trans Evol Comput 11(6):712–731. https://doi.org/10.1109/tevc.2007.892759
21. Zhou A, Qu BY, Li H et al (2011) Multiobjective evolutionary algorithms: a survey of the state of the art. Swarm Evol Comput 1:32–49. https://doi.org/10.1016/j.swevo.2011.03.001

22. Zitzler E, Thiele L (1999) Multiobjective evolutionary algorithms: a comparative case study and the strength Pareto approach. IEEE Trans Evol Comput 3:257–271. https://doi.org/10. 1109/4235.797969
23. Fonseca CM, Fleming PJ (1996) On the performance assessment and comparison of stochastic multiobjective optimizers. Springer, Berlin, Heidelberg
24. López-Ibáñez M, Paquete L, Stützle T (2010) Exploratory analysis of stochastic local search algorithms in biobjective optimization. Experimental methods for the analysis of optimization algorithms. Springer, Berlin, Heidelberg, pp 209–222

Chapter 4
Image Processing

4.1 Introduction

Artificial vision or computer vision aims to provide computer systems with "eyes" to generate a computable representation of the real world. The human vision system provides a large input of information from the environment, allowing to perform high-level tasks such as navigation or decision making. On most mammals, the vision systems are rather complex. As a result, the difficulty to recreate such task on electronic devices has led to different approximations with different degrees of success. The images are analyzed by algorithms that allow decisions to be made according to what exists in that representation. The use of image processing systems allows tasks such as image classification, object tracking, reconstruction, segmentation and feature extraction among many others. However, the scope of the research reflected in this book is limited to the analysis of segmentation techniques.

The set of operations and techniques commonly applied to digital images is known as image processing. Commonly the term vision is used to refer to such techniques. However, there is a subtle difference between image processing and computer vision [1]. For example, image processing techniques are not limited to the visible range of the electromagnetic spectrum, while vision systems are limited to replicating visual processes in images visible to the human eye. Thus, image processing tools can be applied to any type of image; for example, thermal images, electromagnetic resonances, gamma rays, among many others. In any case, the techniques of image processing and vision require as the basis of their processes some representation of the physical world in the form of a digital image. Digital images are captured by a sampling process on an analogue signal from digital capture devices.

Computationally, visual perception systems are constructed with five modules described in Fig. 4.1. First, the images are generated by a specialized sensor and a numeric representation is stored in memory. Then, the preprocessing step is used to improve the quality of the image and make successive steps easier. The third step is the segmentation where pixels are grouped according to similar properties to identify

© Springer Nature Switzerland AG 2019

D. Oliva et al., *Metaheuristic Algorithms for Image Segmentation:*
Theory and Applications, Studies in Computational Intelligence 825,
https://doi.org/10.1007/978-3-030-12931-6_4

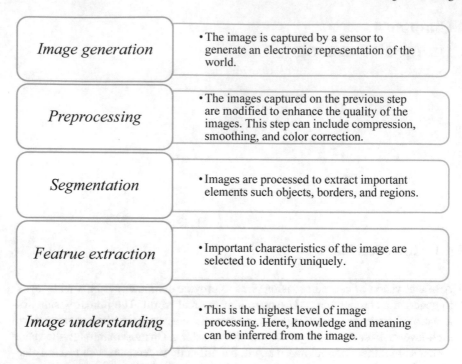

Fig. 4.1 Artificial vision general scheme

objects. Afterward, the extraction of features can be used for higher level tasks on the following step. Finally, the vision system uses the information retrieved on previous steps to "make sense" from the image. This can be done as a classification task or pattern recognition.

4.2   Image Acquisition and Representation

Digital images are generated by an electronic sensor which is excited by the received light. Therefore, a source of light is required to shine over a scene in order to be captured. Then, the light is reflected by the objects of the scene and later acquired by a photosensitive sensor. The material used for the sensor determinates the light wave from the electromagnetic spectrum which is captured. The difference between the different light intensities determinates the content of the image. The most common type of sensor used for capturing digital images is the Charged Coupled Devices (CCD) which are constructed using a matrix of photosites (photosensitive elements). Each photosite generates a voltage when is excited and is recorded as a numeric value. The combination of the spatial localization of the photosensitive element and

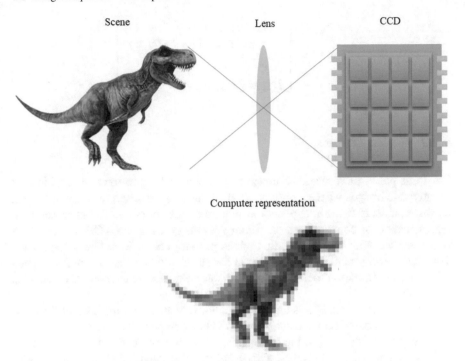

Scene

Lens

CCD

Computer representation

Fig. 4.2 Image acquisition process

its intensity is used to generate a matrix of intensities of the picture, where every element is called pixel (picture element).

Each CCD has its specifications, but two characteristics are the often considered: resolution and depth. The number of photosensitive elements contained in the sensor corresponds to the resolution of the images; a historically relevant resolution is VGA (Video Graphics Array) with 640 by 480 pixels. The other relevant specification is the depth of the image, which indicates the amount of memory used to encode each pixel, with 8-bit images being the most representative example.

The process of acquiring a digital image necessarily includes the processes of capture, sampling, quantification, and coding. Figure 4.2 presents the image acquisition of a scene graphically until its computable form. The most generic form of representation of an image is through a two-dimensional function that quantifies the radiation intensity levels of a portion of the electromagnetic spectrum at a given spatial position. In general, it is expressed as the function $I(u, v)$ where the intensity value is obtained by indexing the spatial positions u and v. Each spatial position is called a pixel, and there is a maximum of N columns and M rows. To facilitate its operation, a matrix-based model is used, as defined in Eq. (4.1).

$$I(u, v) = \begin{bmatrix} I(1,1) & I(2,1) & \cdots & I(N,1) \\ I(1,2) & I(2,2) & \cdots & I(N,2) \\ \vdots & \vdots & \ddots & \vdots \\ I(1,M) & I(2,M) & \cdots & I(N,M) \end{bmatrix} \tag{4.1}$$

4.2.1 Color Images

Until this point, the process of image acquisition and representation is addressed as grayscale images where a bi-dimensional matrix can be used to store the image. However, most real-world applications require the use of color. The most common representation of color images on digital systems is using the RGB model which is constructed with the combination of the primary colors Red, Green, and Blue. This color representation is fundamental for electronic devices as it was designed following the limitations of capturing and display devices such as cameras, scanners, and LCDs.

The RGB representation is an additive color model, which means that the color combination is based on the addition of individual components based on black. This process can be imagined as the overlapping of three beams of red, green and blue light, which are directed towards a dark surface. The intensity of the distinct color components determines both the tone and the resulting color illumination. White and grey or shades of grey are produced equally through the combination of the corresponding three primary RGB colors. On Fig. 4.3 it is possible to observe each axis of the color space.

Each possible color $\mathbf{C}_i$ is generated by the RGB model within a cube generated by the combination of the three colors over the three axes (R, G, B) on the interval $[0, V_{max}]$.

$$\mathbf{C}_i = (R_i, G_i, B_i), 0 \leq R_i, G_i, B_i \leq V_{max} \tag{4.2}$$

The possible intensities of each color are usually normalized between [0,1]. Following this notion, the color black will be $\mathbf{K} = (0, 0, 0)$ while white is $\mathbf{W} = (1, 1, 1)$. However, it is also possible to consider the depth of the image; on 8-bit images, the interval is [0,255].

In Fig. 4.4 the combination of the intensities of three colors is presented following the additive model of the RGB model. Those images are encoded using an 8-bit representation. From a practical point of view, a color image can be represented as a collection of three grayscale images on the RGB color model. Although RGB is enough for most applications, there are other color spaces with interesting properties such as the HSV, HSL, and CMYK.

Every pixel on a true color image can take any value within the selected color space. On RGB, the stored images are generated as a result of a composition of the three planes that correspond to each one of the axes of the color model. By

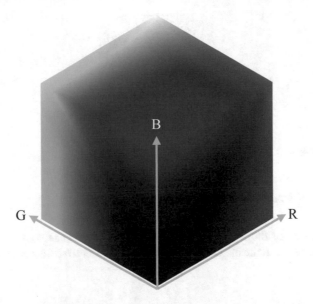

Fig. 4.3 Cube representation of the RGB color space

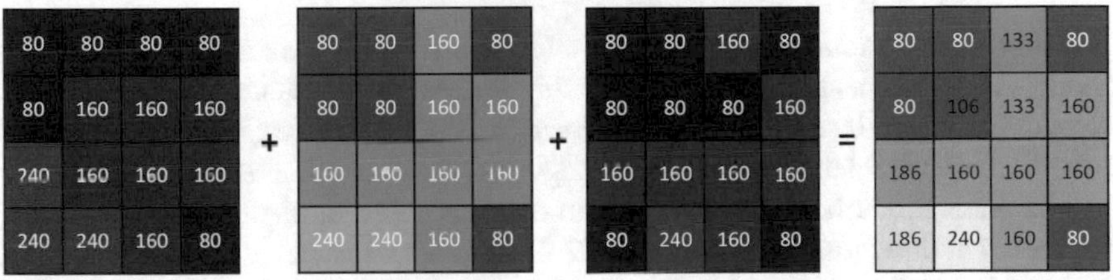

80	80	80	80
80	160	160	160
240	160	160	160
240	240	160	80

+

80	80	160	80
80	80	160	160
160	160	160	160
240	240	160	80

+

80	80	160	80
80	80	80	160
160	160	160	160
80	240	160	80

=

80	80	133	80
80	106	133	160
186	160	160	160
186	240	160	80

Fig. 4.4 Additive model of RGB (The numbers indicate the intensity value of each color R, G, and B, while the resulting images show the averaged result on an 8-bit image)

composition, it is understood that each pixel has a spatial correspondence on every plane and all images arrays are of the same size. Then, a color image is formally described as:

$$I = (I_R, I_G, I_B) \qquad (4.3)$$

where each $I_R(u, v)$, $I_G(u, v)$, and $I_B(u, v)$ contain the intensity values for each color.

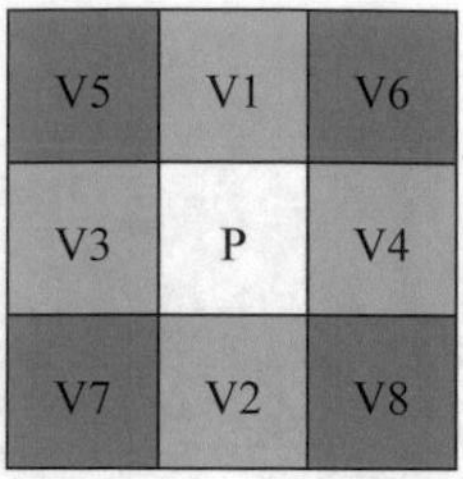

V5	V1	V6
V3	P	V4
V7	V2	V8

Fig. 4.5 Vicinity of pixel P

4.2.2 Pixel Properties

Given the nature of the representation of digital images, it is possible to establish relationships between the elements of the image in order to extract relevant information.

4.2.2.1 Neighborhood or Vicinity

The first of these relationships and of relevance for this book is the vicinity of a pixel. It is defined as the spatial relationship that exists between a pixel and nearby pixels. As the pixels are organized in a grid, it is possible to establish two types of neighborhood; the neighborhood of 4-neighbors and 8-neighbors.

The 4-Neighbourhood is formed from the four adjacent pixels in the up, down, left and right directions (Fig. 4.5, V1, V2, V3 and V4). Instead, the 8-neighborhood is composed of the four pixels of the 4-neighbourhood and the four closest pixels in the diagonal directions (Fig. 4.5, V5, V6, V7 and V8).

4.2.2.2 Connectivity

The definition of vicinity can help to analyze the content of an image by comparing the value of a pixel P with respect with its surrounding. In this way, it is possible to establish connectivity relationships. The connectivity is a widely used on object detection and region segmentation on certain images. The basic idea is grounded on the two basic vicinities. This concept is specially used over binary images where each pixel takes the value of either one or zero. In this sense, two pixels are connected by 4-neighbor connectivity if those two pixels are in a 4-neighborhood vicinity. The same concept applies to the 8-neighborhood connectivity.

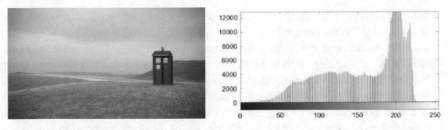

Fig. 4.6 Image and its histogram

4.3 Histogram

4.3.1 Description of Histograms

Histograms are statistical tools that allow us to visualize the distributions described by the frequency with which a set of events occur. From the perspective of image processing, histograms consider as events all possible levels of intensity in the image and each of them is associated with the number of occurrences of that intensity value throughout the image. In the case of grayscale images, an image $I(u, v)$ containing intensity values in the range $[0, L-1]$ will generate an H histogram with exactly L different values. Since in grayscale images it is common to encode the intensities using 8 bits, a total of $L = 2^8 = 256$ intensity levels is obtained. Therefore, the frequency of occurrence of the intensity level i is defined as $h(i)$ and is equal to the number of pixels with intensity i existing in the image I for all values $0 \leq i < L$. Formally defined as:

$$h(i) = card\{(u, v) | I(u, v) = i\} \tag{4.4}$$

where $card\{\ldots\}$ indicates the number of elements. Thus, $h(0)$ it is the number of pixels that have an intensity value of 0. Similarly, $h(1)$ expresses the number of pixels with a value of 1 and thus continues until $h(255)$. The resulting histogram is presented in the form of a one-dimensional vector of length L. In Fig. 4.6 we can see an image next to its histogram.

Histograms, despite their great usefulness, have some limitations. For example, with the histogram we can know how many pixels there are of a certain intensity, but we cannot know the spatial location of these pixels only with the histograms.

4.3.2 Image Characteristics and Enhancement

The histogram is widely used to analyze key features to evaluate the quality of an image, such as contrast and dynamic range and brightness. When an image is captured under poor lighting conditions, the histogram will show such problems, which can be corrected by further processing.

In order to enhance the quality of the captured images, it is possible to modify the images according to their histograms. This is done by performing multiple individual pixel operations designed to reshape the histogram to improve their qualities. Formally, every new pixel $p' = I'(u, v)$ depends on the value of the original image $p = I(u, v)$ at the same position. The new pixel value is determined by a homogeneous function $f(\cdot)$ which is independent of the coordinates of the image:

$$f(I(u, v)) \rightarrow I'(u, v) \tag{4.5}$$

Homogeneous functions can be applied to change contrast and brightness. Also, the shape of the histogram can be modified at will.

4.3.2.1   Brightness

The brightness of a grayscale image is formally described as the average intensity resulting from all pixels of the image. Visually we can observe how bright or dark an image is. Also, the brightness is related to the distribution of intensity values. It is possible to easily modify the brightness of an image by direct addition or subtraction of a constant to each pixel. For example, a homogeneous function which adds 50 to each pixel will generate a great increase on the brightness as:

$$f(I(u, v)) = I(u, v) + 50 \tag{4.6}$$

However, the limits of the image should be considered to avoid saturation (saturation occurs when the values of pixels exceed the possible limits). On Fig. 4.7 we can observe how the artificial modification of the image on Fig. 4.6 generates a darker image (a), and a brighter image (b). Besides, on Fig. 4.7 we can observe how the image is saturated since part of the information exceed the limits.

4.3.2.2   Contrast

The concept of contrast can be simply explained as the difference between the minimum and maximum intensity value present on the image. Following this concept, an image can be said to be of full contrast when it uses intensity values of 0 and L.

The contrast can be easily modified with a homogeneous function which multiplies the value of each pixel by a constant. On Eq. 4.6 the constant used is 1.5 generating a 50% increase in contrast. To visualize the concept of contrast, in Fig. 4.8 the image of Fig. 4.6 is modified to reduce its contrast taking values within the range [50, 150].

$$f(I(u, v)) = I(u, v) * 1.5 \tag{4.7}$$

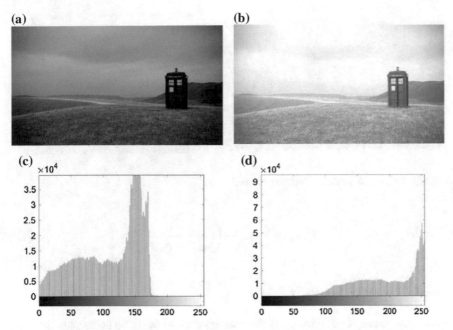

Fig. 4.7 Modified brightness, **a** and **c** the brightness is reduced by 50 while, **b** and **d** brightness is increased by 50

Fig. 4.8 Image with modified contrast

Since the best scenario is when the image is represented with all possible intensity levels there a simple way to automatically enhance the contrast to fit that goal. This technique works by first considering the darker pixel p_{low} as the minimum value while the brightest pixel p_{high} is considered as the maximum value; then, the pixels in-between the minimum and maximum values are linearly interpolated to cover all intensities as described in Eq. 4.8 where L is the maximum possible intensity value and 0 indicates the minimum possible intensity.

Fig. 4.9 Image with modified dynamic contrast

$$f_{ac} = (p - p_{low}) \cdot \left(\frac{L - 0}{p_{high} - p_{low}} \right) \tag{4.8}$$

4.3.2.3 Dynamic Range

The dynamic range of an image is the concept used to describe how many intensity levels are used to represent the image. Ideally, an image is better when it is represented using the complete set of possible levels. The image on Fig. 4.6 has a good dynamic range since its histogram covers most of the gray levels of an 8-bit image. However, on the right side of the histogram of the image, many bright levels are not used. On Fig. 4.9, the image is artificially enhanced to cover all possible intensity values.

4.3.2.4 Cumulative Histogram

The cumulative histogram is a variation of the regular histogram which reflects important information that can be useful for image enhancement; specially to balance the histogram. It is constructed on each level as the sum of the occurrences of previous intensity levels; as a result, it is a monotonic increasing function. An example of the cumulative histogram can be observed in Fig. 4.10 image (e). Formally, the cumulative histogram is computed as:

$$H(i) = \sum_{j=0}^{i} h(j), \text{ for } 0 \leq i \leq L \tag{4.9}$$

The cumulative histogram is used to equalize images in order to obtain better pictures. The process of equalization is defined by Eq. 4.10 where the main idea is to generate a cumulative histogram that grows linearly. In Fig. 4.10 we can observe how the process of equalization modifies the image and both traditional and cumulative histograms.

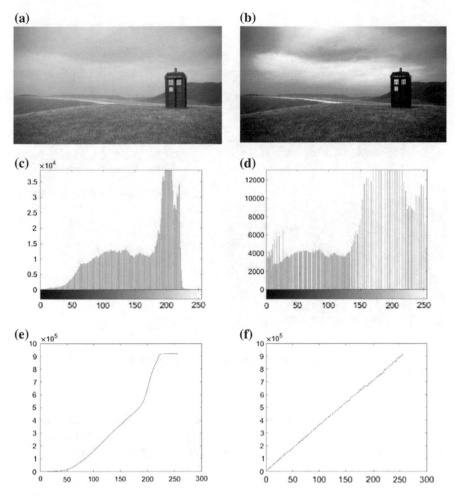

Fig. 4.10 Cumulative histogram an its equalization. **a** Original image, **b** image after equalization, **c** histogram of the original image, **d** histogram of the equalized image, **e** cumulative histogram of the original image, and **f** cumulative histogram of the equalized image

$$f_{eq} = \left(h(p) \cdot \frac{L-1}{M \cdot N} \right) \qquad (4.10)$$

4.3.3 Color Histogram

Since color images are mostly represented as the composition of three color planes, a couple of clarifications should be made. First, each plane or channel has an associated histogram indicating the occurrence frequency of each color intensity as depicted in

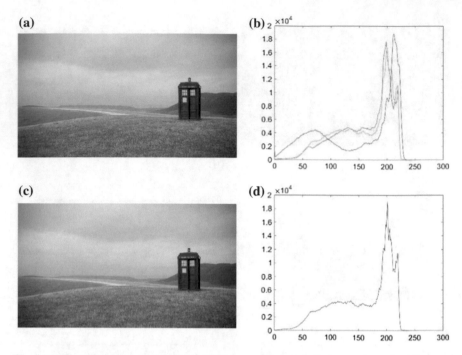

Fig. 4.11 Histogram of color images, **a** and **b** use a histogram for every channel while **c** and **d** present the brightness image or grayscale image

Fig. 4.11. Moreover, the contribution of each pixel on every channel can be averaged to form an intensity or grayscale image. The histogram associated with the grayscale representation of a color image can take the name of brightness histogram.

4.3.4   Binarization

The binarization is the process used to transform an image of intensities into a binary representation. In other words, every pixel will take the values of either zero or one. This process is also called thresholding, and it is the simplest scenario for image segmentation. The binarization occurs after by a logical comparison between the pixel intensity and a threshold value, where the threshold value will separate the image into two classes.

$$f_{th}(p) = \begin{cases} 0 \text{ if } p < th \\ 1 \text{ if } p \geq th \end{cases} \tag{4.11}$$

Fig. 4.12 Binary image
using as $th = 100$

where $0 < th < L$. The values that each pixel can take might be modified according to the application. On Fig. 4.12 we can observe the effect of the binarization of the test image.

4.4 Image Segmentation

In this section, the formal definition of the segmentation of images is presented followed by a brief explanation of the most representative segmentation techniques. A high emphasis is made on image thresholding since it is the main focus of this book. At the end of the description of each method, a short list of pros and cons is presented to summarize the concepts.

4.4.1 Definition of Segmentation

Formally, the problem of segmentation is established as a process that separates an image into representative regions according to at least one criterion that is shared within each region [2]:

Given an image $I(u, v)$ and R its whole representation, according to segmentation, it is possible to state that R contains n subregions R_i, $i = 1, 2, \ldots, n$ such that the following is true:

- The segmentation must be complete. Each pixel contained in the image must be present in a subregion $\bigcup_{i=1}^{n} R_i = R$.
- The pixels contained in R_i share a common property or connectivity.
- The segmentation must be disjunct, $R_i \cap R_j = \phi$ for all i, j where $i \neq j$.
- $P(R_i) = $ True where $P(\ )$ is a logical predicate over the points of the set R_i and the empty set ϕ.
- $P(R_i \cup R_j) = $ False for $i \neq j$.

4.4.2   Segmentation Techniques

Most vision systems require segmentation. Correct segmentation can improve the performance of the entire system in general as it is common for other processes to depend on the results of that segmentation. All segmentation systems can be grouped into two categories: based on homogeneity of regions and based on discontinuity between adjacent regions. Dividing images into regions (partition) allows only regions of interest to be analyzed and processed. In addition, segmentation can be used to reduce memory usage and increase system speed by reducing redundant information.

4.4.2.1   Thresholding

The main objective of segmentation is the partitioning of the image into homogeneous classes, where the elements of each class share common properties such as intensity or texture. Currently, a large number of contributions have been proposed in the literature, which can be found in the following researches [3–5]. One of the simplest segmentation techniques is image thresholding (TH). The TH uses the grayscale histogram to select threshold values that are intended to separate classes in the image histogram. The two-level thresholding is the simplest case and uses only one threshold value to create two classes. Under this approach, an object can be extracted from its background. While the two-level thresholding is easy to implement, the multilevel thresholding (MTH) has more complications, as its goal is to find several classes. Segmentation methods based on thresholding can be divided into parametric and nonparametric [6, 7]. Parametric approaches estimate the parameters of a probability density function to describe each class, but this approach is computationally costly. In contrast, nonparametric approaches use criteria such as class variance, entropy, and error rate to find the best thresholds [8–10]. These criteria are optimized to find the optimal threshold values, providing robust and accurate methods [11]. Otsu proposed a popular method that can find thresholds by maximizing the variance between the classes of foreground and background intensity levels [10]. Otsu's method is considered one of the best threshold selection methods for real-world images [12]. However, the formulation of class variance is ineffective in the case of multilevel thresholds. As the number of levels increases, computational time increases exponentially, and its accuracy decreases with each new threshold point [13]. Another widely used method has been proposed by Kapur [8], and maximizes entropy to measure the homogeneity of each class. Both approaches have been evaluated in various contexts, demonstrating their efficiency and accuracy [13].

In addition, the development of information theory has been adopted to explore the use of various entropies for segmentation purposes. Under these approaches, the objective is to find efficient entropy criteria that with their optimization allow separating regions in an image. Some examples include Kapur entropy, Tsallis entropy [14], and cross entropy [15] to name a few. One of the most important entropy measures is the minimum cross entropy (MCE) criterion [15], which has been widely used in the literature to segment images. This criterion was introduced by Li and Lee in a segmentation algorithm that identifies thresholds minimizing cross entropy between regions of the segmented image. On the other hand, Yin proposed a recursive programming technique that reduces the computation magnitude of the target function of the MCE [16].

As an alternative to parametric techniques, the MTH problem has also been addressed by evolutionary algorithms. Such approaches produced several applications of the TH by selecting various evolutionary computation techniques and optimizing various criteria, such as optimization with particle swarm (PSO) and the firefly algorithm (FF) to use cross entropy as a criterion [17]. Another example is the use of artificial bee colony (ABC) and PSO with Kapur's criterion [6] and finally the use of modified bacterial forage algorithm (MBFO) with Tsallis entropy [18]. Besides, other methodologies have been proposed to incorporate contextual information [19] or treating the TH as a multi-objective problem [20].

Pros:

- Fast and easy to implement.
- No need for external information with automatic approaches.
- Widely used in the literature.

Cons:

- Spatial information is not considered on most versions.
- Noisy images can provide suboptimal results.

4.4.2.2 Edge-Based Methods

This model of segmentation is based on the extraction of edges that are used to identify border regions that separate two or more adjacent classes. They work by locating pixels where the first derivative of the intensity is greater than a predefined threshold or where the second derivative has no crossings. After the identification of the pixel borders using operators like Sobel or Canny, a processing step is required to guarantee the connectivity of the boundaries to limit each region. Then, a binary image is generated as output providing structural information [21].

Pros:

- Sharp and clear edges.
- Good contrast between objects.

Cons:

- False edges can hinder the quality of the segmentation.
- The presence of many edges generates poor results.

4.4.2.3   Region-Based Methods

The idea of segmenting the image into regions according to a shared characteristic is present on all segmentation forms. However, the so-called region-based segmentation techniques include region growing methods or merging and splitting methods [22, 23]. First, region growing segmentation is carried out selecting an initial pixel to be used as a seed. According to the application the seed selection can be automatic or selected automatically. The seed pixel then grows according to the properties of the connectivity on its neighborhood until the algorithm determines its stop.

In the case of region splitting and merging, two basic operators are used; merging and splitting. The split operator iteratively divides an image into regions with similar properties while the merge operator is devoted to combining similar adjacent regions [24].

Pros:

• Can handle well the presence of noise.
• Good contrast between objects.

Cons:

• Requires a greater computer power.
• Consumes more memory than previous techniques.

4.4.2.4   Clustering

Clustering methods are designed to classify into each cluster the available pixels of the image according to a certain characteristic. The basic idea is to group the pixels that are more similar. The clustering process usually takes two forms: by hierarchy or partition. The first models the image as a tree where the root encodes the entire set of internal nodes that describe the clusters. The second form, optimization methods iteratively minimize a fitness function. Those two approaches require different algorithms to find the clusters with two basic types; hard and soft clustering [25, 26].

Hard clustering approaches are known to classify every pixel into only one cluster. The membership functions involved in hard clustering can take values of 1 or 0. The most representative approach is k-means clustering, where all centers of the clusters are determined, then, each pixel is grouped into the closest center by emphasizing the intracluster similarity. The other approach called soft clustering is more relaxed. Here, each pixel can partially belong to multiple clusters. This feature allows to have a better representation of real-life problems, and it is more robust. The main example is fuzzy c-means clustering, where pixels are grouped into the cluster with a degree of membership.

Pros:

• Can handle real-life images with the fuzzy incorporation.

Cons:

- The selection of membership function can harm the performance.

4.4.2.5 Artificial Neural Networks

The Artificial Neural Networks are universal approximators that can be trained to predict or classify a series of data. The basic idea is to emulate the architecture of many neural connections in the brain. Each connection has assigned a weight that enhances or suppress a stimulus. The correct selection of an architecture and weights leads to outstanding classifiers that perform well on image segmentation, especially with a large number of layers [27, 28].

Pros:

- Easy to implement.
- Can be accelerated by some form of parallelization.
- Deep Learning approaches are state of the art on many applications.

Cons:

- The training can take a lot of time and memory.
- The accuracy is highly dependent on the training examples.
- Manual labeling of examples is required in many cases.

4.5 Summary

In this chapter, the basic theory of computer vision and image processing is presented to provide the reader with the required tools to understand the rest of the book. First, the process of image acquisition is described and its computable representation. Then, the basic properties of the pixels are discussed. Since most of the content of this book relies on the use of the histogram of the image, an entire section is dedicated to discussing the properties of the histogram and how it can be modified to enhance the quality of the images; the focus resides on contrast and brightness operations and histogram equalization. Besides, the special case of color images is presented. Finally, the problem of segmentation is formally stated, and the five most representative methodologies for image segmentation are described while highlighting the pros and cons of each method.

References

1. Davies ER, Computer Vision (Fifth Edition), Academic Press (2018) ISBN 9780128092842, https://doi.org/10.1016/B978-0-12-809284-2.00030-7.   (http://www.sciencedirect.com/science/article/pii/B9780128092842000307)
2. Gonzalez RC, Woods RE (1992) Digital image processing. Pearson, Prentice-Hall, New Jersey
3. Osuna-Enciso V, Cuevas E, Sossa H (2013) A comparison of nature inspired algorithms for multi-threshold image segmentation. Expert Syst Appl 40:1213–1219. https://doi.org/10.1016/j.eswa.2012.08.017
4. Zaitoun NM, Aqel MJ (2015) Survey on image segmentation techniques. Procedia Comput Sci 65:797–806. https://doi.org/10.1016/j.procs.2015.09.027
5. Zhang YJ (1996) A survey on evaluation methods for image segmentation. Pattern Recognit 29:1335–1346. https://doi.org/10.1016/0031-3203(95)00169-7
6. Akay B (2013) Bahriye: a study on particle swarm optimization and artificial bee colony algorithms for multilevel thresholding. Appl Soft Comput 13:3066–3091. https://doi.org/10.1016/j.asoc.2012.03.072
7. Hammouche K, Diaf M, Siarry P (2010) A comparative study of various meta-heuristic techniques applied to the multilevel thresholding problem. Eng Appl Artif Intell 23:676–688. https://doi.org/10.1016/j.engappai.2009.09.011
8. Kapur JN, Sahoo PK, Wong AKC (195) A new method for gray-level picture thresholding using the entropy of the histogram. Comput Vis Graph. Image Process 29:273–285. https://doi.org/10.1016/0734-189x(85)90125-2
9. Kittler J, Illingworth J (1986) Minimum error thresholding. Pattern Recognit 19:41–47. https://doi.org/10.1016/0031-3203(86)90030-0
10. Otsu N (1979) A threshold selection method from gray-level histograms. IEEE Trans Syst Man Cybern 9:62–66. https://doi.org/10.1109/TSMC.1979.4310076
11. Sezgin M, Sankur B (2004) Survey over image thresholding techniques and quantitative performance evaluation. J Electron Imaging 13(1), https://doi.org/10.1117/1.1631315
12. Sahoo P, Soltani S, Wong AK (1988) A survey of thresholding techniques. Comput Vis Graph Image Process 41:233–260 (1988). https://doi.org/10.1016/0734-189x(88)90022-9
13. Sathya PD, Kayalvizhi R (2011) Optimal multilevel thresholding using bacterial foraging algorithm. Expert Syst Appl 38:15549–15564. https://doi.org/10.1016/j.eswa.2011.06.004
14. de Albuquerque MP, Esquef IA (2004) Gesualdi Mello AR (2004) Image thresholding using Tsallis entropy. Pattern Recognit Lett 25:1059–1065. https://doi.org/10.1016/j.patrec.2004.03.003
15. Li CH, Lee CK (1993) Minimum cross entropy thresholding. Pattern Recognit 26:617–625. https://doi.org/10.1016/0031-3203(93)90115-D
16. Yin P-Y (2007) Multilevel minimum cross entropy threshold selection based on particle swarm optimization. Appl Math Comput 184:503–513. https://doi.org/10.1016/j.amc.2006.06.057
17. Horng M-H (2011) Multilevel thresholding selection based on the artificial bee colony algorithm for image segmentation. Expert Syst Appl 38:13785–13791. https://doi.org/10.1016/j.eswa.2011.04.180
18. Tang K, Xiao X, Wu J, Yang J, Luo L (2017) An improved multilevel thresholding approach based modified bacterial foraging optimization. Appl Intell 46:214–226. https://doi.org/10.1007/s10489-016-0832-9
19. Oliva D, Hinojosa S, Elaziz MA, Ortega-Sánchez N (2018) Context based image segmentation using antlion optimization and sine cosine algorithm. Multimed. Tools Appl (2018). https://doi.org/10.1007/s11042-018-5815-x
20. Hinojosa S, Avalos O, Oliva D, Cuevas E, Pajares G, Zaldivar D, Gálvez J (2018) Unassisted thresholding based on multi-objective evolutionary algorithms. Knowl Based Syst (2018). https://doi.org/10.1016/j.knosys.2018.06.028
21. Arbeláez P, Maire M, Fowlkes C, Malik J (2011) Contour detection and hierarchical image segmentation. IEEE Trans Pattern Anal Mach Intell 33:898–916. https://doi.org/10.1109/TPAMI.2010.161

22. Angelina S, Suresh LP, Veni SHK (2012) Image segmentation based on genetic algorithm for region growth and region merging. In: 2012 international conference on computing, electronics and electrical technologies (ICCEET), pp 970–974. IEEE

23. Siddiqui AM, Ghafoor A, Khokher MR (2013) Image segmentation using multilevel graph cuts and graph development using fuzzy rule-based system. IET Image Process 7:201–211. https://doi.org/10.1049/iet-ipr.2012.0082

24. Kaganami HG, Beiji Z (2009) Region-based segmentation versus edge detection. In: IEEE, 2009 fifth international conference on intelligent information hiding and multimedia signal processing, pp 1217–1221

25. Chuang K-S, Tzeng H-L, Chen S, Wu J, Chen T-J (2006) Fuzzy c-means clustering with spatial information for image segmentation. Comput Med Imaging Graph 30:9–15. https://doi.org/10.1016/J.COMPMEDIMAG.2005.10.001

26. Dhanachandra N, Manglem K, Chanu YJ (2015) Image segmentation using K-means clustering algorithm and subtractive clustering algorithm. Procedia Comput Sci 54:764–771. https://doi.org/10.1016/J.PROCS.2015.06.090

27. Hall LO, Bensaid AM, Clarke LP, Velthuizen RP, Silbiger MS, Bezdek JC (1992) A comparison of neural network and fuzzy clustering techniques in segmenting magnetic resonance images of the brain. IEEE Trans Neural Netw 3:672–682. https://doi.org/10.1109/72.159057

28. Schmidhuber J (2015) Deep learning in neural networks: an overview. Neural Netw 61:85–117. https://doi.org/10.1016/J.NEUNET.2014.09.003

Chapter 5
Image Segmentation Using Metaheuristics

5.1 Introduction

Image segmentation (IS) [1] is considered as a process of partitioning a digital image into several segments or pixels. Segmentation aims to make a simple representation of the image which has a strong meaning and can be analyzed easier. Also, IS can assign a label for each pixel in the image, where the pixels with the same label share certain characteristics. IS has the ability to locate objects and boundaries such as lines or curves in the images. The output of the IS is a set of segments that can cover the entire image. In the last decades, IS became a point of attention as it can be used in solving many problems such as object recognition, occlusion boundary estimation within motion or stereo systems, image compression, image editing, or image database look-up, and more else.

There is another kind of image segmentation is called color image segmentation which can divide a chromatic image into homogeneous and connected regions based on color, texture and their combination [2]. It is an essential step in image analysis and can determine the final output of the analysis of an image. Segmentation methods can be divided into several categories based on the method which is used like edge, region, fuzzy neural networks based, and feature clustering based or swarm intelligence based.

Many segmentation methods [1] have been brought forward to deal with image segmentation start from traditional method to metaheuristics and swarm intelligence techniques. For example, authors in [3] performed segmentation using the chaotic particle swarm optimization and artificial bee colony optimization, and 2-D Fisher criteria thresholding. The authors in [4] proposed the graph partitioning scheme Swarm Intelligence based Graph Partitioning (SIBGP) for Image Segmentation using the particle swarm optimization [5] for segmenting images.

Mandal in [6] proposed a robust version of the Chan and Vese algorithm which is expected to achieve satisfactory segmentation performance. Irrespective of the initial choice of the contour where he formulated the fitting energy minimization

© Springer Nature Switzerland AG 2019
D. Oliva et al., *Metaheuristic Algorithms for Image Segmentation: Theory and Applications*, Studies in Computational Intelligence 825, https://doi.org/10.1007/978-3-030-12931-6_5

problem to be solved using a metaheuristic optimization algorithm and makes a successful implementation of the algorithm using particle swarm optimization (PSO) technique. Li in [7] proposed that the particles in the swarm are constructed, and the swarm search strategy is used to meet the needs of the segmentation application. Then the fuzzy entropy image segmentation based on particle swarm optimization is implemented, and the proposed method obtains satisfactory results in the segmentation experiments. Osuna-Enciso et al., in [8] proposed a bio-inspired algorithm, which utilizes the allostatic mechanism as a base model which used the Allostatic Optimization (AO) algorithm and made a comparison between it and the Differential Evolution (DE), Artificial Bee Colony (ABC) and Particle Swarm Optimization (PSO). Nebti in [9] proposed using several swarm-based methods the predator-prey optimizer (PPO), the symbiotic algorithm, the cooperative co-evolutionary optimizer and the bees' algorithm for color image segmentation.

In this chapter, we provide a review of three of the most popular image segmentation methods and their improvement by using the meta-heuristic methods. In general, there are various traditional image segmentation techniques which try to find a suitable solution to the image segmentation problem. The most popular three methods are including Edge segmentation, Region Segmentation, and Data Clustering. Where the Edge based segmentation aims to find the boundaries of each object within the image based on some criteria. Also, the Region based image segmentation depends on collect the most similar pixels in regions, meanwhile, the data clustering deals with the pixels of the image as points that required to grouped in a set of classes. Therefore, the basic information about each method of those three image segmentation is introduced in this chapter and give some recommendations in future to work these image segmentation methods.

The rest of the chapter is organized as follows. In Sect. 5.2, the standard Edge segmentation method is introduced. Section 5.3 presents the Region growing methods. Section 5.4 introduces the clustering image segmentation method. Section 5.5 discusses the advantage and disadvantages of each one of the three segmentation methods. Finally, in Sect. 5.6 the conclusions and future recommendations are discussed.

5.2 Edge-Based Image Segmentation

In this section, the image segmentation methods based on edge detection are discussed. In general, the edge detection (ED) methods aim to determine the boundaries between two regions of the same image which separated them based on their grey level characteristics. Therefore, the ED methods have been applied in several image processing applications, for example, Detection of blood vessels [10].

There are several operators applied to determine the suitable edge of the objects such as Prewitt operator, canny operator, Sobel operator, Laplacian, Wallis operator and Kirsch operator. Each operator has its mask which used for edge detection by scaling down the column of data and this leads to remove the irrelevant features and preserve only the relevant features.

Based on these concepts, there are many works have been presented. For example: In [11], the morphological operator is combined with the region growing method, where the morphological is used to improve the quality of the image then the dilation residue edge detector is used. Then they deployed growing seeds and used the region growing process for image segmentation. In [12], the snake boundary condition is applied to obtain get better results of edge detection. The Phase symmetry is applied in [13] to segmentation of prostate boundaries using ultrasound images. Also, the Median filter is used to reduce the noise effect. In addition, the authors used the edge extraction and edge linking to generates the final edge-based segmentation image. Moreover, there are various algorithms used to detect edges such as curve fitting [14], optimization of a criterion [15], statistical testing [16] and soft computing [17].

However, most of the previously mentioned edge detection methods depend on using an $n \times n$ mask that convoluted with the image. With the small size of this mask the time computational is decreased. However, this refers to the limited amount of information that extracted from the area under the mask. To avoid this situation that can affect the performance of the edge detection methods, the all edge pattern must be considered. But in this case, the computation time will be increased. Therefore, a global optimization algorithm is needed for exploring a large space to consider the global structure of the edges to reduce broken edges in a reasonable time.

The meta-heuristic (MH) algorithms are global optimization techniques which aim to find the global optimal solution of the given problem. The MH algorithms, in general, simulates the behaviors of animals, physical, biological. For example, PSO, SSO, WOA, GWO, CSA, BBO, and several others. Based on these behaviors they are applied to improve the edge detection results. For example, Rezaee provides an edge detection method based on ant colony optimization (ACO) by considering the edge of the image as the food source for ants [18].

Gonzalez et al., proposed edge detection method using the optimization of a fuzzy edge detector [19]. This method depends on the combination between the traditional Sobel operator and interval type-2 fuzzy logic. The main aim of using interval type-2 fuzzy logic is to handle the uncertainty in processing real-world images. But the fuzzy system has different parameters that need to be selected in the suitable method. Therefore, the Cuckoo Search (CS) and Genetic Algorithms (GAs) are also used.

In [20], the ACO is presented as an edge detection method, which employs a new heuristic function, adopts a user-defined threshold in pheromone update process and provides a group of suitable parameter values. The improved Teaching Learning Based Optimization (TLO) algorithm is used for determining the edge maps of the real noisy images [21]. Also, the Harmony Search Algorithm (HSA) is used to detect the circular shapes automatically in images [22].

5.3 Region-Based Image Segmentation

In this kind of image segmentation methods, the images are split into a different number of regions that constructed by dissociating or associating their neighbor pixels [23]. These methods collect the pixels based on the homogeneity, where the neighboring pixels inside a region have the same characteristics, and they are dissimilar with the pixels of the other regions. These collected pixels representing an object which group together. In general, there are several properties such as grey level, shape, texture, and color. Based on this concepts, the region based segmentation methods can provide good results than the edge detection method since they are comparatively manageable and less sensitive to noise [24]. There are several regions based segmentation techniques that have been introduced, for example, region growing, region merging and splitting, etc.

5.3.1 Region Growing

It is an image segmentation method in which the neighboring pixels are explored and joined into a region that represents a group where edges are not identified [25]. This process is repeated for the pixel of the boundary in the region, and if the adjacent regions are determined, then these regions are merged using a suitable method. In this case, the strong edges are maintained, and the other weak edges are removed. The region growing methods can be classified into two categories; the first category is the Seeded Region Growing (SRG) that considered as a semiautomatic method. The SRG is free parameters method, rapid, and robust. The process begins by choosing the seed pixel in which the selection process of seed is very important which has the largest effect on the quality of the segmented image [26]. The second category, which called Unseeded Region Growing (UsRG) that considered as a fully automatic method [27]. The UsRG depends on the similarity between the pixel of the region, and this method is flexible and it is based on tuning parameters.

5.3.2 Region Split and Merge Method

This kind of region-based image segmentation is different from the region growing method which worked on the complete image [28]. Where, it consists of two main phases region split and region merge, so it is called region split-merge (RSM) method. In the region split, that considered as a top-down method; the image is split into several different sliced which are more homogenous than the whole image. However, this process, splitting, is not sufficient for image segmentation since it limits the segmented shapes. Therefore, the second phase that called region merging is used. In general, the RSM method depends on the quadtree which aims to distinguish the homogeneity of the image. In this method, the given image is represented as a single region and then using a specified criterion to split the image into four quadrants.

However, these region-based models are suffering from some limitations that start from select the initial region to the complexity and time computational. So, the meta-heuristic methods are used to solve reduce the effect of these limitations. For example, authors in [29] proposed two MH methods for improving the segmentation of river mapping and flood evaluation according to the multi-temporal time-series analysis of satellite images. Also, they used the spectral information of the image for image clustering and region based segmentation for extracting water covered regions. These two methods are the Particle Swarm Optimization (PSO) and Genetic Algorithm (GA) which aims to distinguish between the water regions and the other non-water regions based on spectral information.

In [30], the PSO is combined with the region-based image segmentation method called seed region growing. Where the PSO aims to solve the two kinds of problems, faced the seed region growing such the process selecting the similarity criteria of the pixels within the region. Meanwhile, the second problem is the process of determining the seeds. The authors in [31] proposed two versions of SRG to determine the multi-seeded region growing problem, using the Particle Swarm Optimization to solve the localization problem.

In [32], a color image segmentation based on an alternative meta-heuristic method called cuckoo search (CS) is used. In general, the CS simulates the behaviors of the cuckoo search in natural [32], and these behaviors are converted to an optimization method to find the global optimal value. So, the CS algorithm is used to optimize the seeds that selected depends on the threshold method. Then regions are grown from the initial seed point by combining neighboring pixels that are similar concerning the intensity level. The authors in [33] introduced a segmentation method to improving the segmentation process of medical Magnetic Resonance Image (MRI). This method is a hybrid between the genetic algorithm and the seed region growing method where the GA aims to find the optimal seeds that used by the region growing algorithm to determine accurate region. In addition, this method provides a new fitness function to evaluate the agents that contain two constraints namely, covers and uncovers penalty.

In [34], to overcome the problem of selecting the seed pixel in the region based segmentation the Genetic-based Fuzzy Seeded Region Growing Segmentation (GFS-RGS) algorithm is used. The GFSRGS optimizes the choosing process of multiple seed pixels using GA based fuzzy method. Also, the authors in [35] proposed the GA with fuzzy initialization and seeded modified region growing (GFSMRG) method with back propagation neural network (BPNN).

5.4 Data Clustering-Based Image Segmentation

Data clustering is a process whereby a data set is replaced by clusters, that can be considered as collections of data points that "belong together." In such case, we can think of image segmentation as clustering; we would desire to represent an image regarding clusters of pixels that "belong together." The application can be considered as the criterion to be used. Pixels may belong together because they have the same color, or the same texture, or the same shape, or they are nearby, etc.

Currently, there are many clustering methods for the large-scale images segmentation which can process the images before doing clustering, and this can be performed with a less complicated clustering algorithm. Also, there are many algorithms for performing such clustering task like watershed spectral clustering [36], level set algorithm [37], and K-means clustering algorithm [38] and more else. Also, there are two kinds of simple clustering algorithms [39], divisive clustering, in which, the entire data set is considered as a cluster, and then clusters are recursively divided to yield a good clustering. Agglomerative clustering, each data item is considered as a cluster and clusters are recursively merged to yield a good clustering. In addition, clustering algorithms may be classified as Exclusive Clustering, Overlapping Clustering, Hierarchical Clustering, and Probabilistic Clustering.

In recent years, the image segmentation problem was solved efficiently based on the data clustering techniques. For example, the K-means clustering algorithm is provided as a clustering based segmentation in [40] to cluster the integrating region and boundary information. In [41], the authors used the combination between the K-means and the Edde detection for color image segmentation. Also, the Hill-climbing is combined with K-means algorithm for color image segmentation [42].

Moreover, the Fuzzy c-means (FCM) clustering algorithm is used for image segmentation in several kinds of literature since it depends on the fuzzy concept that avoids the limitations presented in the K-mean algorithm. In which by using the FCM, the object can belong to more than one class at the same time. AN example from these FCM based image segmentation methods, the clustering based on the fuzzy ant system pixel that extracts different features such as gradient value, grayscale value, and pixel neighborhood information, and used them for color image segmentation [43]. Tan [44] introduced a hybrid between the region split and merge method with FCM. This method is called Region Splitting and Merging-Fuzzy C-means Hybrid Algorithm (RFHA), and it applied to color segmentation. The authors in [45] presented a segmentation method for the 3-D magnetic resonance (MR) images; this method is an adaptive spatial FCM clustering algorithm. Moreover, there are several clustering algorithms such as mean shift clustering [46], and spectral clustering [47] that also applied for image segmentation.

With the advantages achieved by the previous mentioned cluster image segmentation methods, they still require improvements since they can suffer from some limitations such as easily stuck in local point due to the image segmentation is considered as a multi-modal optimization problem. Therefore, the global optimization methods like meta-heuristic algorithms are used alone or combined with clustering method. For example, in [48], a hybrid artificial fish swarm algorithm (HAFSA) which combines artificial fish swarm algorithm (AFSA) with Fuzzy C Means (FCM) for artificial grid graph and Magnetic Resonance Imaging (MRI). Das in [49] proposed an algorithm for fuzzy dynamic clustering called AFDE algorithm for the problem of image segmentation based on the Differential Evolution (DE) approach.

Yu et al. [50] introduced a new ant colony optimization ACO based fuzzy clustering algorithm FCM, and it was applied in image segmentation field. The authors in [51] had employed the method of metaheuristic search known as the Harmony Search algorithm (HS) for generating the near-optimal initial cluster centers for the FCM algorithm called HSFCM. Alomoush in [52] proposed a fuzzy clustering algorithm using hybrid firefly algorithm FA with fuzzy c-means algorithm FCM which is called FFCM to produce a new segmentation method for real and simulated MRI brain images.

The grey wolf optimization (GWO) is applied for satellite image segmentation in which the GOW is used as the clustering method [53]. Also, the hybrid between the dynamic PSO and K-means are proposed as image segmentation method as in [54]. The hybrid between the Differential Evolution, Particle Swarm Optimization, and Fuzzy C-Means Clustering is applied as image segmentation algorithm.

5.5 Comparison Between the Traditional Image Segmentation Methods

In this section, the advantages of each method are given and their limitations as given in Table 5.1. In addition, these methods are compared based on several parameters such as multiple object detection, spatial information, speed, noise resistance, automaticity, accuracy, computation complexity, and region continuity is performed (Table 5.2).

5.6 Conclusions

In recent years, the image segmentation techniques have more attention since they are used in several applications such as medical image, industrial, computer vision, and many other applications. In general, there are many image segmentation methods which include the edge based segmentation, clustering based segmentation, and the region growing segmentation. Each of them has its advantages and disadvantages properties. Therefore, this chapter provides a review of these three image segmentation methods and their extension by meta-heuristic methods that try to improve the limitations of each segmentation method where the MH methods are a global optimization that aims to find the global solution and avoid stagnation to local optimal point through a competitive selection between the individual of the population. This leads to improve the performance of the segmented method and also improve the quality of the segmented image.

Based on the review discussed in this paper there are introduced some recommendations that must be taken into consideration in the future such as:

Table 5.1 The advantages and limitations of each traditional image segmentation method.

Method	Advantages	Limitations
Edge-based segmentation	- They provide good results with an image that have good contrast between the background and objects	- They not suitable to deal with images that have low contrast – Sensitive to noise and smooth transitions - It is not trivial for a robust edge
Region growing based segmentation	- They provide good results better in comparison with Edge method - They are more flexible to select between interactive and automatic image segmentation method - Start from inner pixel until the outer region that produced more clear object boundaries - Suitable choose the initial seed generates an accurate result than other segmentation methods	- They are more computational in time and memory than other methods since they are sequential by nature - The process of selecting the terminal condition - The process of randomly select the initial seed can be stuck on the stagnation in the local point
Clustering based segmentation	- The k-means and fuzzy c mean is faster than other segmentation methods when K/c is small - These methods generate more homogeneous regions than other methods, and FCM gives better than K-means - FCM is an unsupervised method - Removes the noisy spots	- The process of determine K/c is hard and in some cases it is considered NP-hard problem - In addition, they are sensitive to initial value of the center and cluster number which effect on the final quality - For FCM is very hard to determine the fuzzy membership by using traditional method - Most of these method like K-mean/FCM are computationally expensive when the objects number increased

Table 5.2   The comparison between the algorithms based on several parameters.

Parameter	Edge-based method	Region-based method	Cluster-based method
Spatial information	Ignored	Considered	Considered
Region continuity	Reasonable	Good	Reasonable
Speed	Slow	Fast	Moderate
Computation complexity	Moderate	Rapid	Rapid
Automaticity	Interactive	Semiauto	Automatic
Noise resistance	Less	Less	Moderate
Multiple object detection	Poor	Fair	Fair
Accuracy	Moderate	Fine	Moderate

1. It is believed that the image segmentation approach still required to be improved and applied in a new application and different kind of images such as galaxy image.
2. Develop the image segmentation method to be become more automated/semi-automated and reduce the influence of human in this process.
3. Improve the MH methods by considering the combination of different methods to benefit from the properties of each method.
4. Combine the MH with new recently deep neural network techniques and applied these new techniques for image segmentation.

References

1. Unit MI (1993) A review on image segmentation techniques. Pattern Recognit 26:1277–1294
2. Ilea DE, Whelan PF (2011) Image segmentation based on the integration of colour – texture descriptors—a review 44:2479–2501. https://doi.org/10.1016/j.patcog.2011.03.005
3. Chen H (2012) Image segmentation using thresholding and swarm intelligence. J Softw 7:1074–1082. https://doi.org/10.4304/jsw.7.5.1074-1082
4. Dhanya GS, Raj RJS, Sivasamy P (2017) Image segmentation using swarm intelligence based graph partitioning 2017:17–22
5. Kennedy J, Eberhart R (1995) Particle swarm optimization. In: IEEE International Conference on Particle Swarm Optimization, vol 4, pp 1942–1948
6. Mandal D, Chatterjee A, Maitra M (2014) Engineering Applications of Artificial Intelligence Robust medical image segmentation using particle swarm optimization aided level set based global fitting energy active contour approach. Eng Appl Artif Intell 35:199–214. https://doi.org/10.1016/j.engappai.2014.07.001
7. Li L, Li D (2008) Fuzzy entropy image segmentation based on particle swarm optimization. Prog Nat Sci 18:1167–1172. https://doi.org/10.1016/j.pnsc.2008.03.020
8. Osuna-Enciso V (2014) Bioinspired metaheuristics for image segmentation. ELCVIA Electron Lett Comput Vis Image Anal 13:1–3

9. Nebti S (2010) Color image segmentation using swarm based optimisation methods. LNCS, vol 6377, pp 277–284
10. Gökhan M (2009) Detection of blood vessels in ophthalmoscope images using MF/ant (matched filter/ant colony) algorithm 6:85–95. https://doi.org/10.1016/j.cmpb.2009.04.005
11. Detection ME (2005) A contour based image segmentation algorithm using morphological edge detection. In: 2005 IEEE International Conference on Systems, Man and Cybernetics, pp 2962–2967
12. Haider W, Malik MS, Raza M, Wahab A (2012) A hybrid method for edge continuity based on pixel neighbors pattern analysis (PNPA) for remote sensing satellite images. Int J Commun Netw Syst Sci 5:624–630. https://doi.org/10.4236/ijcns.2012.529072
13. Zaim A (2008) An edge-based approach for segmentation of prostate ultrasouind. In: 3rd International Symposium on Communications, Control and Signal Processing, ISCCSP 2008. IEEE, pp 12–14
14. Chen G, Yang YHH (1995) Edge detection by regularized cubic b-spline fitting. IEEE Trans Syst Man Cybern 25:636–643
15. Nascimento JC, Marques JS (2005) Adaptive snakes using the EM algorithm. IEEE Trans Image Process 14:1678–1686
16. Lim DH (2006) Robust edge detection in noisy images. Comput Stat Data Anal 50:803–812. https://doi.org/10.1016/j.csda.2004.10.005
17. Alper B, Enis G (2009) Efficient edge detection in digital images using a cellular neural network optimized by differential evolution algorithm. Expert Syst Appl 36:2645–2650. https://doi.org/10.1016/j.eswa.2008.01.082
18. Rezaee A (2008) Communications 59:57–59
19. Gonzalez CI, Melin P, Castro JR et al (2016) Optimization of interval type-2 fuzzy systems for image edge detection. Appl Soft Comput J 47:631–643. https://doi.org/10.1016/j.asoc.2014.12.010
20. Banharnsakun A (2018) Artificial bee colony algorithm for enhancing image edge detection. In: Evolving Systems, pp. 1–9. https://doi.org/10.1007/s12530-018-9255-7
21. Thirumavalavan S, Jayaraman S (2016) An improved teaching – learning based robust edge detection algorithm for noisy images. J Adv Res 7:979–989. https://doi.org/10.1016/j.jare.2016.04.002
22. Ortega-sánchez ECN, Pérez-cisneros DZM (2012) Circle detection by harmony search optimization, 359–376. https://doi.org/10.1007/s10846-011-9611-3
23. Chondro P, Yao C, Ruan S, Chien L (2017) Low order adaptive region growing for lung segmentation on plain chest radiographs. Neurocomputing 0:1–10. https://doi.org/10.1016/j.neucom.2017.09.053
24. Zeng Y, Liao S, Tang P et al (2018) Automatic liver vessel segmentation using 3D region growing and hybrid active contour model. Comput Biol Med 97:63–73. https://doi.org/10.1016/j.compbiomed.2018.04.014
25. Region U, Pulse G, Neural C et al (2018) Unsupervised color image segmentation with color-alone feature using region growing pulse coupled neural network. Neurocomputing 306:1–16. https://doi.org/10.1016/j.neucom.2018.04.010
26. Kang C, Wang W, Kang C (2012) Image segmentation with complicated background by using seeded region growing. AEUE Int J Electron Commun 66:767–771. https://doi.org/10.1016/j.aeue.2012.01.011
27. Lin Z, Jin J, Talbot H (2000) Unseeded region growing for 3D image segmentation, Selected p. Australian Computer Society, Inc.
28. Wang Z, Jensen JR, Im J (2010) Environmental Modelling & Software An automatic region-based image segmentation algorithm for remote sensing applications. Environ Model Softw 25:1149–1165. https://doi.org/10.1016/j.envsoft.2010.03.019
29. Senthilnath J, Shenoy HV, Omkar SN, Mani V (2013) Spectral-spatial MODIS image analysis using swarm intelligence algorithms and region based segmentation for flood assessment, pp 163–174. https://doi.org/10.1007/978-81-322-1041-2

30. Mohsen F, Hadhoud M, Mostafa K, Amin K (2012) A new image segmentation method based on particle swarm optimization. Int Arab J Inf Technol 9:487–493

31. Mirghasemi S, Rayudu R, Zhang M (2013) A new image segmentation algorithm based on modified seeded region growing and particle swarm optimization. In: 2013 28th International Conference on Image and Vision Computing New Zealand, pp 382–387

32. Synthuja MM, Preetha J, Suresh LP, Bosco MJ (2015) Cuckoo search based color image segmentation using seeded region growing, pp 1573–1583. https://doi.org/10.1007/978-81-322-2119-7

33. Zanaty EA, Ghiduk AS (2013) A novel approach based on genetic algorithms and region growing for magnetic resonance image (MRI) segmentation. ComSIS 10. https://doi.org/10.2298/CSIS120604050Z

34. Tamilarasi M (2013) Genetic based fuzzy seeded region growing segmentation for diabetic retinopathy images

35. Kavitha AR, Chellamuthu C, Kavitha AR (2016) Brain tumour segmentation from MRI image using genetic algorithm with fuzzy initialisation and seeded modified region growing (GFSMRG) method. 2199. https://doi.org/10.1080/13682199.2016.1178412

36. Liu L, Wen X, Gao X, Technology NS (2010) Segmentation for SAR image based on a new spectral clustering algorithm, pp 635–643

37. Yazdi MB, Khalilzadeh MM, Foroughipour M (2014) Fuzzy c-means clustering method based on prior knowledge for brain MR image segmentation, pp 235–240

38. Wu M, Lin C (2015) Brain tumor detection using color-based K-means clustering segmentation brain tumor detection using color-based K-means clustering segmentation, pp 1–5. https://doi.org/10.1109/IIHMSP.2007.4457697

39. Mesejo P, Ibáñez Ó, Cordón Ó, Cagnoni S (2016) A survey on image segmentation using metaheuristic-based deformable models: State of the art and critical analysis. Appl Soft Comput J 44:1–29. https://doi.org/10.1016/j.asoc.2016.03.004

40. Oliver A, Mu X, Batlle J (2006) Improving clustering algorithms for image segmentation using contour and region information, pp 1–6

41. Bala A, Sharma AK (2016) Color image segmentation using k-means clustering and Morphological Edge Detection Algorithm. Int J Latest Trends Eng Technol 48–55

42. Chandana BS, Srinivas K, Kumar RK (2014) Clustering algorithm combined with hill climbing for classification of remote sensing image. Int J Electr Comput Eng (IJECE) 4:923–930

43. Han Y, Shi P (2007) An improved ant colony algorithm for fuzzy clustering in image segmentation. 70:665–671. https://doi.org/10.1016/j.neucom.2006.10.022

44. Tan KS, Ashidi N, Isa M, Lim WH (2017) Color image segmentation using adaptive unsupervised clustering approach. Appl Soft Comput J 13:2017–2036. https://doi.org/10.1016/j.asoc.2012.11.038

45. Liew AW, Yan H, Member S (2003) An adaptive spatial fuzzy clustering algorithm for 3-D MR image segmentation. IEEE Trans Med Imaging 22:1063–1075

46. Tao W, Jin H, Member S (2007) Color image segmentation based on mean shift and normalized cuts. IEEE Trans Syst Man Cybern Part B (Cybern) 37:1382–1389

47. Tilton JC (1998) Image segmentation by region growing and spectral clustering with a natural convergence criterion. Inst Electr Electron INC 4:1766–1768

48. Ma L, Li Y, Fan S, Fan R (2015) A hybrid method for image segmentation based on artificial fish swarm algorithm and fuzzy-means clustering 2015. https://doi.org/10.1155/2015/120495

49. Das S, Konar A (2009) Automatic image pixel clustering with an improved differential evolution 9:226–236. https://doi.org/10.1016/j.asoc.2007.12.008

50. Yu Z, Yu W, Zou R, Yu S (2009) On ACO-based fuzzy clustering for image segmentation. In: Yu W, He H, Zhang N (eds) Springer, Heidelberg

51. Moh O, Mandava R, Ramachandram D, Aziz ME (2009) Dynamic fuzzy clustering using harmony search with application to image segmentation, pp 538–543

52. Alsmadi M (2014) A hybrid firefly algorithm with fuzzy-c mean algorithm for MRI brain segmentation. Am J Appl Sci 11:1676–1691. https://doi.org/10.3844/ajassp.2014.1676.1691

53. Kapoor S, Zeya I, Singhal C, Nanda SJ (2017) A grey wolf optimizer based automatic clustering algorithm for satellite image segmentation. Procedia Comput Sci 115:415–422. https://doi.org/10.1016/j.procs.2017.09.100
54. Li H, He H, Wen Y (2015) Dynamic particle swarm optimization and K-means clustering algorithm for image segmentation. Opt Int J Light Electron Opt 126:4817–4822. https://doi.org/10.1016/j.ijleo.2015.09.127

Chapter 6
Multilevel Thresholding for Image Segmentation Based on Metaheuristic Algorithms

6.1 Introduction

Currently, image thresholding continues attracting the attention of researchers from different areas, due to the growing demand for computer vision systems over the last decade. Digital cameras are ubiquitous nowadays; multiple electronic devices are equipped with cameras and require specific software for image treatment and understanding for applications such as surveillance, medical diagnosis, industrial implementations, etc. In this context, the first stage of this kind of systems is the segmentation [1].

Thresholding (TH) is a specific form of segmentation which consists in the separation of the pixels into different groups depending on their intensity level according to one or more threshold values [2]. Thresholding is not only easy but also is robust and it can work with the noisy images. Based on the number of thresholds required for the image, the TH process can be divided into two categories: (1) bi-level and (2) multilevel. In the bi-level approach, the pixels that have an intensity value higher than the threshold are labeled as an object, and the remaining pixels are part of the background. On the other hand, the multilevel TH (MTH) uses several thresholds to separate the pixels in different regions that represent objects contained in the image. For the segmentation of real-life images, the best alternative is the use of multilevel thresholding [3].

The thresholding problem can be summarized as the search for the best threshold values on an image. It should be noticed that the threshold points depend on the histogram of the image; thus, each image has its own set of best threshold values [3]. The well-known methods to select the best thresholds are Otsu's method [4] and Kapur's method [5]. However, the searching for the best thresholds in MTH is a NP-hard problem, and it was determined as a challenge in several studies [3, 6, 7]. If there is a small number of thresholds, classical methods are suitable to solve this problem, otherwise, the segmentation's accuracy will be affected, and it will be computationally expensive and the computing time for exhaustive search

© Springer Nature Switzerland AG 2019
D. Oliva et al., *Metaheuristic Algorithms for Image Segmentation:
Theory and Applications*, Studies in Computational Intelligence 825,
https://doi.org/10.1007/978-3-030-12931-6_6

increases exponentially regard to the required thresholds. So, to solve these problems, metaheuristic algorithm (MA) and swarm intelligence methods (SI) are combined with thresholding methods.

MA are good search methods that applied to solving several complex problems such as [8–13]. They simulate the behaviors of animals or birds in the natural in order to enhance the solutions during the iterative process. There are many types of MA such as Particle Swarm Optimization (PSO) [14], Ant Colony Optimization (ACO) [15], Firefly Algorithm (FA) [16], and Social Spider Optimization (SSO) [17].

Some MA are already applied for MTH segmentation problems. They are used to optimize the objective function of the problem such as Otsu's and Kapur's methods. They used different types of images and thresholds' levels Several studies applied these methods such as Wind Driven Optimization (WDO) [18], SSO [7], Honey Bee Mating Optimization (HBMO) [19], Harmony Search (HS) algorithm [20], Bacterial Foraging Optimization (BFO) [21], Cuckoo Search (CS) [22], and Artificial Bee Colony (ABC) [23]. Therefore, in this chapter, we will present a survey of the most used metaheuristics algorithms that used in solving MTH segmentation problem.

This chapter is arranged as follow. Section 2 defines the multilevel thresholding segmentation problem. Section 3 presents the studies of metaheuristics algorithms in multilevel thresholding image segmentation. The conclusion is listed in the last section.

6.2   Multilevel Thresholding

This section presents the problem definition of MTH. It is mathematically represented by assuming an image I is a gray level and consists of $K + 1$ classes. So, K thresholds are needed to split an image into subparts. The following equation explains this process [24]:

$$C_0 = \{g(i, j) \in I \,|\, 0 \le g(i, j) \le t_1 - 1\}$$
$$C_1 = \{g(i, j) \in I \,|\, t_1 \le g(i, j) \le t_2 - 1\}$$
$$C_2 = \{g(i, j) \in I \,|\, t_2 \le g(i, j) \le t_3 - 1\}$$
$$C_3 = \{g(i, j) \in I \,|\, t_3 \le g(i, j) \le t_4 - 1\}$$

$$\vdots$$

$$\vdots$$

$$C_k = \{g(i, j) \in I \,|\, t_k \le g(i, j) \le L - 1\} \tag{6.1}$$

where C_k defines the k-th class of the current image, t_k $(k = 1, \ldots K)$ defines the threshold value. $g(i, j)$ and L represent the gray level of (i, j) and I, respectively, in the range $(0, 1, \ldots, L - 1)$.

Therefore, the main aim of MTH is to find the threshold values that divide pixels into subparts.

6.3 Meta-Heuristic Algorithms for Segmentation

This section introduces several studies that used the bio-inspired algorithm in solving the MTH segmentation problem. these algorithms are: cuckoo search algorithm, bat algorithm, artificial bee colony, firefly algorithm, social-spider optimization, whale optimization algorithm, moth-flame optimization, and gray wolf algorithm as well as some hybrid methods.

6.3.1 Cuckoo Search Algorithm

Cuckoo search algorithm (CS) was introduced by Yang and Deb [25]. It simulates the behavior of the obligate brood parasitism of some kind of cuckoo. They put their eggs in the nests of other host birds. These eggs may be discovered or not by the birds based on their colors and patterns; the best nests and eggs represent the solutions in an optimization process. Several studies use CS in solving MTH problems such as the authors of [26] used CS to optimize Kapur method for MTH to segment ten satellite images using four level of thresholds. ELR-CS and WDO were also used for comparison. The performance of the CS was measured using fitness value, PSNR, SSIM, and FSIM. The experiment revealed that the three algorithms could be accurately and efficiently applied in MTH problem.

In the same trend, the authors of [27] tried to segment three satellite images using MTH over four levels of thresholds namely 5, 7, 9, and, 11. The CS was compared with two modified versions of CS; the first one modified the flight generation in CS algorithm by McCulloch's method and the second one was modified by Mantegna's methods as well as it was compared with PSO, Darwinian Particle Swarm Optimization (DPSO), and ABC algorithms. The objective functions were Kapur method, Otsu's method, and Tsallis entropy. MSE, PSNR, FSIM and CPU time was used as a performance measure. The improved version of CS by McCulloch's method evolved to be most promising.

In [28] the authors compared CS with FF, PSO, and DE in optimizing Kapur's and Otsu's methods for segmenting six images over four levels of thresholds. The results showed that, CS outperformed the other algorithms except for FF in thresholds number 4 and 5. In addition, In [29] the search methodology of CS algorithm was modified by each of Lévy Flight (LF), Chaotic search, and Brownian Distribution (BD). Otsu's method was used as an objective function, whereas, RMSE, PSNR, and SSIM were used as performance measures. Five benchmark images and four thresholds' levels were applied. The experiment showed that, chaotic and BD based-CS provided better objective value, SSIM, and PSNR, whereas LF based-CS showed faster convergence.

6.3.2   Bat Algorithm

Bat algorithm (BAT) was proposed by Yang [30]; it simulates the echo ranging behavior of real bats to perform an optimization process. It showed a good performance in solving many optimization problems. Therefore, it used in several MTH segmentation studies. In [31] Bat algorithm was used to optimize the fuzzy entropy method and applied to solve MTH segmentation problem on eight natural and infrared images over four levels of thresholds. It efficiency converged to optimal threshold compared with GA, PSO, ACO, and ABC as well as BAT based on Kapur and Otsu's methods.

In [32] the BAT was improved by chaotic theory to optimize Otsu's method. Sixteen benchmark images and four thresholds' levels were tested. The CPU time, a fitness value, PSNR, SSIM, and RMSE were calculated to check the performance of the algorithm. The results showed a good performance of the improved-BA against the compared methods.

An improved version of BAT was also applied to segment medical images using Otsu's method. The improvement used the modified random localization strategy to increase the efficiency of bat search as well as the convergence speed. The algorithm was tested using two medical images and three level of thresholds; its results outperformed the compared algorithms [33].

Also, the BAT was also improved in [29] and employed to optimize Kaper and Otsu's method to solve MTH segmentation problem. It was tested on six benchmark images and four levels of thresholds. The results showed the superiority of the improved BAT especially in a large number of thresholds compared with PSO, CS, DE, and FF in term of fitness value and CPU time.

6.3.3   Artificial Bee Colony

The Artificial Bee Colony (ABC) proposed by Karaboga [34]. It simulates the behavior of the honeybees' colony. It employs three kinds of bees, the worker, onlooker, and scout bees' groups; all these groups work together to search for new sources of food. In the optimization model, the food source is determined as a solution [34].

This algorithm was applied in several studies to optimize the MTH segmentation problem such as the authors of [35] applied the MTH using Kapur's method based-ABC as a pre-segmentation phase to detect the iris. The experiments used only 2 and 3 thresholds, and the results of the ABC and G-best-ABC showed good performance when comparing with Cuckoo Search (CS) and PSO algorithms.

Also in [36], the ABC was improved using a pool of optimal foraging strategies to escape from getting trap in local optima. The improved ABC used multi-dimensional PSO to combine local search and comprehensive learning. The experiment tested six images over three thresholds using Kapur's method and the results of the segmentation showed a significant improvement of the proposed method.

Moreover, in [37] the ABC improved by merging multi population cooperative mechanism and bee-to-bee communication pattern. This algorithm used to optimize Otsu's method and was successfully tested in solving MTH problems using six benchmark images and one MRI image over six levels of thresholds. ABC was modified and employed for MTH segmentation of satellite images in [38] The experiment applied three fitness functions namely Kapur's, Otsu's methods, and Tsallis entropy as well as five images and four thresholds were used. The result of the proposed method was compared with PSO and GA and showed good performance in PSNR, SSIM, MSE, and CPU time.

6.3.4   Firefly Algorithm

Firefly Algorithm (FF) was introduced by Yang [39]. It based on the Firefly community that uses the brightness of Firefly as a strategy to attract the less bright one; that means, the less bright firefly will fly toward the most brighter. Based on this concept, the FF algorithm determines the optimal solution based on the brightness and attractiveness as well as it updates both of them constantly. Several studies used this algorithm in MTH segmentation such as in [40] FF was used and compared with SSO and the results showed that the FF slight outperformed SSO in all measures.

In addition, the authors of [28] compared FF with CS, PSO, and DE in optimizing Kapur's and Otsu's methods for segmenting six images over four levels of thresholds. The results showed that FF outperformed the other algorithms in thresholds number 4 and 5 as well as it achieved a lower computation time.

In [41] the FF was guided by Brownian distribution (BD) to perform MTH segmentation and compared with the classic FF and FF guided by Lévy flight (LF) over 12 test images. The BD-FF showed the best results in term of better objective function, SSIM, and PSNR, whereas LF-FF showed faster convergence and lower computation time.

Furthermore, the author of [42] the FF was improved by two strategies namely neighborhood strategy and diversity-enhancing with Cauchy mutation; then it was employed for improving the efficiency of MTH segmentation using Otsu's method. The improved FF was tested over seven images, and its results showed better performance then Darwinian PSO, hybrid DE, and classic FF.

6.3.5   Social-Spider Optimization

Social-Spider Optimization (SSO) algorithm is a type of swarm optimization approach introduced by Cuevas in [17]. SSO simulates the natural behavior of social spiders to optimize a given problem. The solution of a problem is represented as a spiders' position in a spider web that is used for transferring the information between spiders. The spiders communicate with each other using the vibration that done when they move to a new position in the web.

The spiders' positions are changed based on the gender of them due to the male and female have different strategies in changing the positions since the females represent more than 75% of the spiders' community. The optimal solution is calculated as an offspring output that is produced by mating the dominant males with neighbor female, and the result is compared with the previous optimal solution, and the best one is preserved. SSO was used for MTH segmentation in several studies as follow.

The authors of [7] SSO was applied to improve the Kapur's and Otsu's methods. Five images were tested, and the results showed that, the SSO outperformed the PSO in objective function value and CPU time. In [43], the SSO optimized Kapur's and Otsu's methods and was compared with flower pollination, PSO, and BAT algorithms in term of SSIM, PSNR, fitness function, and CPU time. The experiments recorded that, the SSO achieved the best results in all measures except for CPU time and showed a good balance between exploitation and exploration process.

The SSO was also applied in [44] and compared with firefly algorithm (FF). The SSO act similar to FF in the tested image with a slight superiority to FF; therefore, the authors of [40] was combined SSO with FF for MTH and produced good results and short segmentation time against the-state-of-arts algorithms.

6.3.6 Whale Optimization Algorithm

Whale Optimization Algorithm (WOA) [45] is a metaheuristic algorithm which mimics the social behavior of humpback whales. As in [46] Abd El Aziz proposed an efficient method for determining the multilevel thresholding values for image segmentation. This method considered the multilevel threshold as a multi-objective function problem and used the WOA to solve this problem. Mostafa et al. in [47] proposed using the WOA for liver segmentation in MRI images based which is used to extract the different clusters in the abdominal image to support the segmentation process.

6.3.7 Moth-Flame Optimization

Moth-Flame Optimization (MFO) [48] is also a metaheuristic algorithm which emulates the navigation mechanism of moths in nature. For example, Muangkote in [49] proposed an improved version of the MFO algorithm for image segmentation to effectively enhance the optimal multilevel thresholding of satellite images. As well as in [3] the authors implemented MFO for MTH and compared it with WOA; the results showed good performance of MFO.

6.3.8 Grey Wolf Optimization

Grey Wolf Optimization (GWO) [50] is an algorithm inspired by gray wolves' leadership and hunting behaviors to solve optimal reactive power dispatch (ORPD) problem. As in [51], Li introduced fuzzy Kapur's entropy as the optimal objective function, with modified discrete GWO as the tool, uses pseudo-trapezoid-shaped to conduct fuzzy membership initialization to achieve image segmentation finally using local information aggregation.

Li in [52] put forward the modified discrete grey wolf optimizer algorithm (MDGWO), which improved on the optimal solution updating mechanism of the search agent by the weights. Taking Kapur's entropy as the optimized function and based on the discreteness of threshold in image segmentation. KOC in [53] proposed the grey wolf optimizer (GWO), a recently swarm-based meta-heuristic which imitates the social leadership and hunting behavior of gray wolves in nature employed for solving the multilevel image thresholding problem.

6.3.9 Particle Swarm Optimization

Particle Swarm Optimization (PSO) [54] algorithm, for example, Maitra in [55] presented an optimal multilevel thresholding algorithm for histogram-based image segmentation by using an improved variant of PSO, a relatively recently introduced stochastic optimization strategy. Ghamisi in [56] proposed a new multilevel thresholding method is introduced for the segmentation of hyperspectral and multispectral images which is based on fractional-order Darwinian particle swarm optimization (FODPSO) which exploits the many swarms of test solutions that may exist at any time.

Yeng Yinin [57] first presented a recursive programming technique which reduces an order of magnitude for computing the MCET objective function. Then, a particle swarm optimization (PSO) algorithm was proposed for searching the near-optimal MCET thresholds. Gao in [58] proposed the quantum-behaved PSO employing the cooperative method (CQPSO) to save computation time and to conquer the curse of dimensionality by preserving the fast convergence rate of particle swarm optimization (PSO) for image segmentation based on the Multilevel thresholding. Feng in [59] proposed the image thresholding approach based on the index of entropy maximization of the 2-D grayscale histogram to deal with the infrared image. The threshold is obtained by the Particle Swarm Optimization (PSO) algorithm. Li in [60] proposed a dynamic-context for cooperative quantum-behaved particle swarm optimization algorithm for multilevel thresholding applied to medical image segmentation.

In addition to more researchers made hybrids from these algorithms, for example, Abd El Aziz in [3] proved the ability of two nature-inspired algorithms namely: Whale Optimization Algorithm (WOA) and Moth-Flame Optimization (MFO) to determine the optimal multilevel thresholding for image segmentation. In addition, in [61] the WOA was combined with PSO to improve the segmentation phase; Otsu's and fuzzy entropy methods were applied. Six images over 6 levels of thresholds were tested. Akay in [62] proposed two successful swarm-intelligence-based global optimization algorithms, particle swarm optimization (PSO) and artificial bee colony (ABC), which were employed to find the optimal multilevel thresholds.

6.4 Conclusions

Multilevel image thresholding is a method applied to segment images. It usually employed in image preprocessing phase. In this chapter, a survey of using meta-heuristics in multilevel image segmentation was presented. Nine algorithms were listed namely cuckoo search, bat algorithm, artificial bee colony, particle swarm optimization, firefly algorithm, social spider optimization algorithm, whale optimization algorithm, moth-flame optimization algorithm, and the gray wolf optimization algorithm, as well as their studies in multilevel image segmentation, were described. From these studies, we can conclude that, the most used objective functions are Otsu and Kapur. Whereas, the most used performance measures are the value of the objective function, CPU time, PSNR, and SSIM. The average of test images are 6 images and the average of the levels of thresholds are four namely 2, 3, 4, and 5. In future, we will apply MTH to segment color images.

References

1. Sezgin M, Sankur B (2004) Survey over image thresholding techniques and quantitative performance evaluation. J Electron Imaging 13:146–168. https://doi.org/10.1117/1.1631316
2. Dirami A, Hammouche K, Diaf M, Siarry P (2013) Fast multilevel thresholding for image segmentation through a multiphase level set method. Sig Process 93:139–153
3. El Aziz MA, Ewees AA, Hassanien AE (2017) Whale optimization algorithm and moth-flame optimization for multilevel thresholding image segmentation. Expert Syst Appl 83:242–256. https://doi.org/10.1016/j.eswa.2017.04.023
4. Otsu N (1979) A threshold selection method from gray-level histograms. IEEE Trans Syst Man Cybern 9:62–66. https://doi.org/10.1109/TSMC.1979.4310076
5. Kapur JN, Sahoo PK, Wong AK (1985) A new method for gray-level picture thresholding using the entropy of the histogram. Comput vision, Graph image Process 29:273–285
6. Marciniak A, Kowal M, Filipczuk Paweł and Korbicz J (2014) Swarm intelligence algorithms for multi-level image thresholding. In: Intelligent Systems in Technical and Medical Diagnostics. Springer, pp 301–311
7. Agarwal P, Singh R, Kumar, Sandeep Bhattacharya M (2016) Social spider algorithm employed multi-level thresholding segmentation approach. Proceedings of First International Conference on Information and Communication
8. Elaziz MEA, Ewees AA, Oliva D, et al (2017) A hybrid method of sine cosine algorithm and differential evolution for feature selection. In: International Conference on Neural Information Processing, pp 145–155

9. Ibrahim RA, Elaziz MA, Ewees AA et al (2018) Galaxy images classification using hybrid brain storm optimization with moth flame optimization. J Astron Telesc Instruments, Syst 4:38001

10. Ewees AA, Elaziz MA, Houssein EH (2018) Improved Grasshopper Optimization Algorithm using Opposition-based Learning. Expert Syst Appl

11. Ibrahim RA, Oliva D, Ewees AA, Lu S (2017) Feature selection based on improved runner-root algorithm using chaotic singer map and opposition-based learning

12. Houssein EH, Ewees AA, ElAziz MA (2018) Improving twin support vector machine based on hybrid swarm optimizer for heartbeat classification. Pattern Recognit Image Anal 28:243–253

13. Ibrahim RA, Elaziz MA, Lu S (2018) Chaotic opposition-based grey-wolf optimization algorithm based on differential evolution and disruption operator for global optimization. Expert Syst Appl 108:1–27

14. Kennedy J, Eberhart RC (1995) Particle swarm optimization. In: 1995 Proceedings IEEE International Conference on Neural Networks, vol 4, pp 1942–1948. https://doi.org/10.1109/icnn.1995.488968

15. Kaveh A, Talatahari S (2010) An improved ant colony optimization for constrained engineering design problems. Eng Comput 27:155–182

16. Yang X-S (2009) Firefly algorithms for multimodal optimization. In: International Symposium on Stochastic Algorithms, pp 169–178

17. Cuevas E, Cienfuegos M, Zald\'\iVar D, PéRez-Cisneros M (2013) A swarm optimization algorithm inspired in the behavior of the social-spider. Expert Syst Appl 40:6374–6384

18. Bayraktar Z, Komurcu M, Bossard JA, Werner DH (2013) The wind driven optimization technique and its application in electromagnetics. IEEE Trans Antennas Propag 61:2745–2757

19. Horng M-H (2010) A multilevel image thresholding using the honey bee mating optimization. Appl Math Comput 215:3302–3310

20. Oliva D, Cuevas E, Pajares G, et al (2013) Multilevel thresholding segmentation based on harmony search optimization. J Appl Math. https://doi.org/10.1155/2013/575414

21. Bakhshali Mohamad Amin, Shamsi M (2014) Segmentation of color lip images by optimal thresholding using bacterial foraging optimization (BFO). J Comput Sci 5:251–257

22. Agrawal S, Panda R, Bhuyan S, Panigrahi BK (2013) Tsallis entropy based optimal multilevel thresholding using cuckoo search algorithm. Swarm Evol Comput 11:16–30. https://doi.org/10.1016/j.swevo.2013.02.001

23. Bhandari AK, Kumar A, Singh GK (2015) Modified artificial bee colony based computationally efficient multilevel thresholding for satellite image segmentation using Kapur's, Otsu and Tsallis functions. Expert Syst Appl 42:1573–1601

24. Sarkar S, Sen N, Kundu A, et al (2013) A differential evolutionary multilevel segmentation of near infra-red images using Renyi's entropy. In: Proceedings of the International Conference on Frontiers of Intelligent Computing: Theory and Applications (FICTA), pp 699–706

25. Yang X-S, Deb S (2014) Cuckoo search: recent advances and applications. Neural Comput Appl 24:169–174

26. Bhandari AK, Singh VK, Kumar A, Singh GK (2014) Cuckoo search algorithm and wind driven optimization based study of satellite image segmentation for multilevel thresholding using Kapur's entropy. Expert Syst Appl 41:3538–3560

27. Suresh S, Lal S (2016) An efficient cuckoo search algorithm based multilevel thresholding for segmentation of satellite images using different objective functions. Expert Syst Appl 58:184–209

28. Brajevic I, Milan T (2014) Cuckoo search and firefly algorithm applied to multilevel image thresholding. In: Cuckoo search and firefly algorithm, pp 115–139

29. Alihodzic A, Tuba M (2014) Improved bat algorithm applied to multilevel image thresholding. Sci World J

30. Yang X-S (2010) A new metaheuristic bat-inspired algorithm. In: Nature inspired cooperative strategies for optimization (NICSO 2010). Springer, pp 65–74

31. Ye ZW, Wang MW, Liu W, Chen SB (2015) Fuzzy entropy based optimal thresholding using bat algorithm. Appl Soft Comput 31:381–395

32. Satapathy SC, Raja NSM, Rajinikanth V, et al (2016) Multi-level image thresholding using Otsu and chaotic bat algorithm. Neural Comput Appl 1–23
33. Zhou G, Zhou Y, Li L, Ma M (2018) Modified bat algorithm with Otsu's method for multilevel thresholding image segmentation. J Comput Theor Nanosci 12:4560–4572. https://doi.org/10.1166/jctn.2015.4401
34. Karaboga D (2005) An idea based on honey bee swarm for numerical optimization
35. Bouaziz A, Draa A, Chikhi S (2015) Artificial bees for multilevel thresholding of iris images. Swarm Evol Comput 21:32–40
36. Gao Y, Li X, Dong M, Li HP (2018) An enhanced artificial bee colony optimizer and its application to multi-level threshold image segmentation. J Cent South Univ 25:107–120
37. Li JY, Zhao YD, Li JH, Liu XJ (2015) Artificial bee colony optimizer with bee-to-bee communication and multipopulation coevolution for multilevel threshold image segmentation. Math Probl Eng
38. Bhandari AK, Kumar A, Singh GK Modified artificial bee colony based computationally efficient multilevel thresholding for satellite image segmentation using Kapur's, Otsu and Tsallis functions. Expert Syst Appl 42:1573–1601
39. Xin-She Yang (2010) Engineering Optimization: An Introduction with Metaheuristic Applications. John Wiley & Sons, Inc
40. El Aziz MA, Ewees AA, Hassanien AE (2016) Hybrid swarms optimization based image segmentation
41. Raja N, Rajinikanth V, Latha K (2014) No Title. Otsu based Optim multilevel image Threshold using firefly algorithm 37
42. Chen K, Zhou Y, Zhang Z, et al (2016) Multilevel image segmentation based on an improved firefly algorithm. Math Probl Eng 2016
43. Ouadfel S, Taleb-Ahmed A (2016) Social spiders optimization and flower pollination algorithm for multilevel image thresholding: a performance study. Expert Syst Appl 55:566–584. https://doi.org/10.1016/j.eswa.2016.02.024
44. Singh R, Agarwal P, Kashyap M, Bhattacharya M (2016) Kapur's and Otsu's based optimal multilevel image thresholding using social spider and firefly algorithm. In: 2016 International Conference on Communication and Signal Processing (ICCSP), pp 2220–2224
45. Mirjalili S, Lewis A (2016) The whale optimization algorithm. Adv Eng Softw 95:51–67. https://doi.org/10.1016/j.advengsoft.2016.01.008
46. El Aziz MA, Ewees AA, Hassanien AE, et al (2018) Multi-objective whale optimization algorithm for multilevel thresholding segmentation
47. Mostafa A, Hassanien AE, Houseni M, Hefny H (2017) Liver segmentation in MRI images based on whale optimization algorithm. Multimed Tools Appl 1–24. https://doi.org/10.1007/s11042-017-4638-5
48. Mirjalili S (2015) Moth-flame optimization algorithm: a novel nature-inspired heuristic paradigm. Knowl -Based Syst 89:228–249. https://doi.org/10.1016/j.knosys.2015.07.006
49. Muangkote N, Sunat K, Chiewchanwattana S (2016) Multilevel thresholding for satellite image segmentation with moth-flame based optimization. In: 2016 13th International Joint Conference on Computer Science and Software Engineering (JCSSE), pp 1–6
50. Mirjalili S, Mirjalili SM, Lewis A (2014) Grey Wolf Optimizer. Adv Eng Softw 69:46–61. https://doi.org/10.1016/j.advengsoft.2013.12.007
51. Li L, Sun L, Kang W et al (2016) Fuzzy multilevel image thresholding based on modified discrete grey wolf optimizer and local information aggregation. IEEE Access 4:6438–6450
52. Li L, Sun L, Guo J, et al (2017) Modified discrete grey wolf optimizer algorithm for multilevel image thresholding. Comput Intell Neurosci 2017. https://doi.org/10.1155/2017/3295769
53. Koc I, Baykan OK, Babaoglu I (2018) Multilevel image thresholding selection based on grey wolf optimizer. J Polytech Derg 21:841–847
54. Poli R, Kennedy J, Blackwell T (2007) Particle swarm optimization. Swarm Intell 1:33–57. https://doi.org/10.1007/s11721-007-0002-0
55. Maitra M, Chatterjee A (2008) A hybrid cooperative-comprehensive learning based PSO algorithm for image segmentation using multilevel thresholding. Expert Syst Appl 34:1341–1350. https://doi.org/10.1016/j.eswa.2007.01.002

56. Ghamisi P, Couceiro MS, Martins FML, Benediktsson JA (2014) Multilevel image segmentation based on fractional-order Darwinian particle swarm optimization. IEEE Trans Geosci Remote Sens 52:2382–2394
57. Yin P-YP (2007) Multilevel minimum cross entropy threshold selection based on particle swarm optimization. Appl Math Comput 184:503–513. https://doi.org/10.1109/SNPD.2007.85
58. Gao H, Xu W, Sun J, Tang Y (2010) Multilevel thresholding for image segmentation through an improved quantum-behaved particle swarm algorithm. IEEE Trans Instrum Meas 59:934–946. https://doi.org/10.1109/TIM.2009.2030931
59. Feng D, Wenkang S, Liangzhou C et al (2005) Infrared image segmentation with 2-D maximum entropy method based on Particle Swarm Optimization (PSO). Pattern Recognit Lett 26:597–603
60. Li Y, Jiao L, Shang R, Stolkin R (2015) Dynamic-context cooperative quantum-behaved particle swarm optimization based on multilevel thresholding applied to medical image segmentation. Inf Sci (Ny) 294:408–422
61. Ewees AA, Elaziz MA, Oliva D (2018) Image segmentation via multilevel thresholding using hybrid optimization algorithms. J Electron Imaging 27:63008
62. Akay B (2013) A study on particle swarm optimization and artificial bee colony algorithms for multilevel thresholding. Appl Soft Comput J 13:3066–3091. https://doi.org/10.1016/j.asoc.2012.03.072

Chapter 7
Otsu's Between Class Variance and the Tree Seed Algorithm

7.1 Introduction

In recent years, the meta-heuristic techniques have more attention since they are used in several applications such as medical, chemistry, and industrial. In addition, the MH methods have been used in many image processing applications such as objects identification, face recognition, computer vision, and others. In these applications, the image segmentation is a necessary step which divided that image into different groups with the same characteristics such as contrast, texture, color, brightness, and grey level) based on a predefined criterion [1]. Based on the importance of image segmentation, it has been used in different applications including satellite image [2], medical diagnosis [3].

There are many methods used to segment the images such as region extraction [4], histogram thresholding, edge detection [5], and clustering algorithms [6]. However, threshold segmentation [7] is considered one the popular methods can be used to segment the images into its parts according to estimate the threshold value [8]. These methods contain two categories; the first one is the bi-level method which grouped the objects of the image into two classed; meanwhile, the second method is called the multi-level which can split the pixels of the image into different classes based on the intensity [6]. The bi-level method is suitable only when the image contains two classes only, however, when the number of classes increased it becomes not suitable. To solve this problem, the multi-level threshold method is used. Several approaches are proposed using the image histogram to find the best thresholds by maximizing or minimizing a fitness function such as Otsu, and Kapur, entropy.

However, the traditional method which used to determine the optimal thresholds needs more computational time, and to avoid this limitation the meta-heuristic methods are used. For example, the firefly optimization algorithm (FA) is used with fuzzy c-means to find the multi-threshold as in [9], and this algorithm is applied to MRI brain images. The differential evolution (DE) has been applied for multilevel thresholding where the minimum cross entropy thresholding (MCET) is used as an

© Springer Nature Switzerland AG 2019
D. Oliva et al., *Metaheuristic Algorithms for Image Segmentation:
Theory and Applications*, Studies in Computational Intelligence 825,
https://doi.org/10.1007/978-3-030-12931-6_7

objective function. The ant colony optimization (ACO) was proposed for image segmentation and provides good results based on the ground truth data. In addition, there are several methods such as PSO [5, 7, 10], honey bee mating optimization (HBMO) [11], artificial bee colony (ABC) [12, 13], cuckoo search (CS) [14], and harmony search (HS) algorithm [15].

In this chapter, we present a new multi-level thresholding approach based on the tree seed for image segmentation. In which the proposed TSA method aims to maximize the Otsu function to find the optimal multi-level thresholding. The proposed method starts by using a set of initial solutions as the population, then evaluate the performance of each solution by computing the Otsu function. Then the next step is determining the best solution based on the high fitness function. After that, the population is updated using the operators of TSA, and these steps are repeated until reached to the terminal conditions.

The organization of the chapter is as follows: in Sect. 7.2, the problem formulation of the multi-level threshold image segmentation is given. Section 7.3 presents a tree seed algorithm for image segmentation. Section 7.4 introduced the results and discussion. The conclusion and the future work are presented in Sect. 7.5.

7.2 Problem Formulation

The multi-level thresholding problem is defined as the process of split the image I into its classes. For example, consider the image has a set of $K + 1$ groups and to perform this process it is necessary to find a set of K threshold values ($t_k, k = 1, 2 \ldots, K$). The mathematical definition of this problem can be formulated as [1]:

$$
\begin{aligned}
C_0 &= \{I_{ij} \in I \mid 0 \leq I_{ij} \leq t_1 - 1\} \\
C_1 &= \{I_{ij} \in I \mid t_1 \leq I_{ij} \leq t_2 - 1\} \\
&\quad \ldots \\
C_K &= \{I_{ij} \in I \mid t_K \leq I_{ij} \leq L - 1\}
\end{aligned}
\tag{7.1}
$$

where L is the gray levels of I. $C_k(k = 1, 2 \ldots, K)$ is the k-th class corresponding to the threshold t_k.

In order to find the set of threshold values the multi-level thresholding problem is considered as an optimization problem which aims to maximize or minimizing an objective function. In this chapter, we consider the maximization of the Otsu's function as [16]:

$$
t_1^*, t_2^*, \ldots, t_K^* = \max_{t_1, t_2, \ldots, t_K} F(t_1, t_2, \ldots, t_K)
\tag{7.2}
$$

where F is the Otsu's function which defined as [16]:

$$F_{Otsu} = \sum_{i=0}^{K} \theta_i \times (\mu_i - \mu_1)^2, \theta_i = \sum_{j=t_i}^{t_{i+1}-1} P_j \qquad (7.3)$$

$$\mu_i = \sum_{j=t_i}^{t_{i+1}-1} \frac{iP_j}{\theta_j}, P_i = \frac{Fr_i}{N_p}, t_0 = 0, t_{K+1} = L$$

where P_i and Fr_i represents the frequency, and the probability of the i-th gray level of I, respectively. The μ_1 and N_p represent the mean intensity and the total number of pixels of I, respectively.

7.3 Tree Seed Algorithm for Image Segmentation

The tree–seed algorithm (TSA) is proposed in [17] as new global optimization method which simulates the relationship between the trees and their seeds. In which, the trees using their seeds to spread to the surface, and then the seeds are grown, and a new tree is generated from these seeds. Therefore, in TSA, the surface represents the search domain, while, the locations of the trees and seeds represent the solutions of the given problem. Where, the searching process for the best solution is controlled by search tendency (ST) parameter that gives the TSA the local intensification and improve the convergence toward, near, the optimal solution of the test problem [17].

Similar to other meta-heuristic algorithms the TSA begins by determining the initial locations of N trees (X) (i.e., the population) using the following equation.

$$X_{ij} = L + rand \times (U - L), i = 1, 2, \ldots, N, j = 1, 2, \ldots, Dim \qquad (7.3)$$

where, L and U represent the lower bound and higher bound of the search domain, respectively, while, the $rand \in [0, 1]$ is a random number. The Dim represents the dimension of the given problem.

The process of generating a new seed S_i is considered the major factor which effect on the performance of TSA where there exists two method to generate the seed. In the first method, the best solution (X^*) is used as defined in the following Eq. [17]:

$$S_{ij} = X_{ij} + \alpha_{ij} \times \left(X_j^* - X_{rj} \right) \qquad (7.4)$$

where α_{ij} is a random number. While, in the second method, the new seed S_i is generated using random tree (solution) X_r as defined in the following equation [17]:

$$S_{ij} = X_{ij} + \alpha_{ij} \times \left(X_{ij} - X_{rj} \right) \qquad (7.5)$$

However, the process of determining which method must be used to generate a new seed is controlled by using the search tendency ($ST \in [0, 1]$) parameter. Where, in the case that the value of ST is higher this mean that the TSA has a fast convergence and powerful local search. While, in the case the, value of ST is low the TSA has powerful global search but with slow convergence.

Also, to control the number of seeds generated from the current tree, 10 and 25% from the population is set as the minimum and maximum number of seeds produced for a tree, respectively.

The tree–seed algorithm (TSA) [17]

Input: Determine the input parameters such as N: size of population, t_{max}: maximum number of iterations, dim: dimension of the given problem (threshold level), ST: parameter of control the new seed.

Generate set of N trees (X) with dim.
Repeat
Compute the objective function for $X_i, i = 1,2, \dots, N$.
For i=1:N
Find the number of seed used to produce X_i
For all seeds
For $j = 1$: dim
IF $r_1 < ST$
  Using Equation (7.3) to update X_{ij}
Else
  Using Equation (7.4) to update X_{ij}
END IF
END For
If best seed is better than X_{ij} then replace X_{ij} otherwise not update X_{ij}
END FOR
Determine X^*
  $t = t + 1$.
Until ($t < t_{max}$)
Return the best solution X^*.

7.4   Experimental and Results

The performance of the proposed method to improve the image segmentation by finding the optimal thresholds is evaluated in this section. Where a set of images are used, also, the results of the proposed method is compared with other algorithms including, SSO, SCA, ABC, and PSO.

7.4.1 Benchmark Image

In this study, a set of the six images from Berkeley University database are used. These images have different characteristics for example, irregular distributions that can observed Fig. 7.1.

7.4.2 Parameter Settings

In order to investigate the efficacy of the proposed TSA method, it compared with other MH methods such as Particle Swarm Optimization (PSO) [18], Sine cosine Algorithm [19–21] and the social spider optimization (SSO) [22, 23]. The common parameters between these algorithms are set as, population size, a maximum number of iterations. Meanwhile, the other parameters for each algorithm are set as an original reference.

All the tested images are evaluated at different levels of threshold such as 2,3, 4 5 10, and 15. And each algorithm is performed 25 times for fair comparison between the algorithms. All experiments were performed using MATLAB 2017b on windows10 (64bit).

7.4.3 Performance Metrics

There are a set of measures are used to evaluate the quality of the segmented image which including objective value, PSNR, SSIM, and CPU time(s).

The Peak-Signal-to-Noise Ratio (PSNR): it is used to compare the similarity of the original image with the segmented image, and it is defined as:

$$PSNR = 20 \log_{10}\left(\frac{255}{RMSE}\right), \text{(dB)}$$

$$RMSE = \sqrt{\frac{\sum_{i=1}^{ro}\sum_{j=1}^{co}(I_{Gr}(i,j) - I_s(i,j))}{ro \times co}} \tag{7.6}$$

From Eq. 7.6 I_{Gr} is the original image, I_s is the segmented image. Meanwhile, ro and co are the maximum number of rows and columns of the image.

The Structure Similarity Index (SSIM) is used to compare the structures of the original segmented image with the segmented image [24], and it is defined as:

$$SSIM(I_{or}, I_{th}) = \frac{(2\mu_{I_{or}}\mu_{I_{th}} + C1)(2\sigma_{I_{or}I_{th}} + C2)}{(\mu_{I_{or}}^2 + \mu_{I_{th}}^2 + C1)(\sigma_{I_{or}}^2 + \sigma_{I_{th}}^2 + C2)}$$

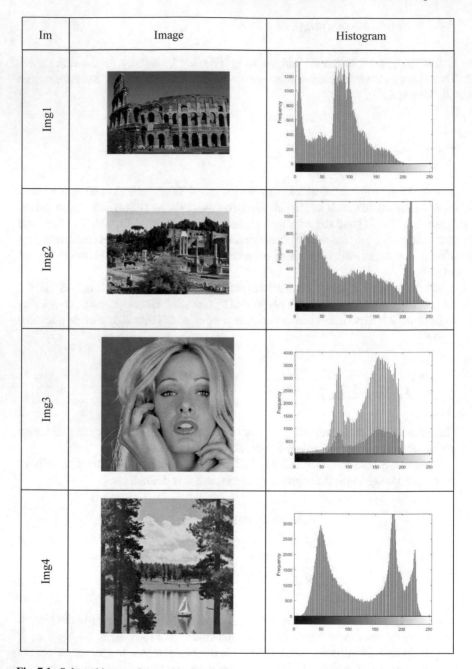

Fig. 7.1 Selected images for graphical analysis

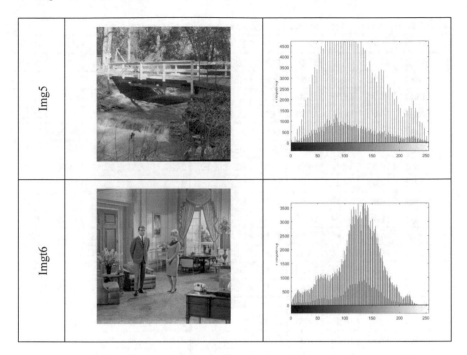

Fig. 7.1 (continued)

$$\sigma_{I_{th}I_{or}} = \frac{1}{N-1} \sum_{i=1}^{N} \left(I_{or_i} + \mu_{I_{or}}\right)\left(I_{th_i} + \mu_{I_{th}}\right) \tag{7.7}$$

where the mean of the original image is $\mu_{I_{or}}$ and the mean of the thresholded image is represented by $\mu_{I_{th}}$. In the same way, for each image, the values of $\sigma_{I_{Gr}}$ and $\sigma_{I_{th}}$ correspond to the standard deviation. $C1$ and $C2$ are constants used to avoid the instability when $\mu_{I_{Gr}}^2 + \mu_{I_{th}}^2 \approx 0$. The values of $C1$ and $C2$ are set to 0.065 considering the experiments of [25].

7.4.4 Results and Discussions

The comparison results of the proposed TSA approach with other approaches are given Tables 7.1, 7.2, 7.3 and 7.4 and Fig. 7.2. Table 7.1 presents the average of the PSNR value at each level for all the algorithms. From this table it can be observed that the proposed TSA method has a higher PSNR value in general. However, nearly, all the algorithms have the same performance at level 2 of the threshold, also, it can see that the PSO gives better results at the threshold 3. But when the level of threshold increase the PSO becomes the worst algorithm, Unlike PSO, the performance of the

Table 7.1 The PSNR results for each algorithm at the different threshold levels

k	Im	ABC	SCA	SSO	PSO	TSA	k	ABC	SCA	SSO	PSO	TSA
2	Img1	14.07	17.35	18.20	17.37	18.37	5	21.64	20.90	20.91	23.84	24.42
	Img2	11.12	14.27	14.20	13.63	12.44		17.74	17.08	19.31	17.17	20.47
	Img3	12.74	14.77	11.05	12.87	12.90		17.43	16.82	17.56	17.05	19.19
	Img4	16.02	15.61	15.74	15.95	15.97		21.12	17.30	19.41	17.59	23.26
	Img5	10.85	12.00	14.53	14.54	14.90		18.31	17.38	17.12	18.98	22.98
	Img6	16.87	14.40	16.45	15.80	16.23		20.84	18.39	16.45	19.27	22.26
3	Img1	18.45	18.77	19.98	20.09	16.63	10	24.12	25.31	23.84	21.08	26.78
	Img2	14.93	15.29	15.15	15.75	16.79		22.22	22.80	23.09	20.62	24.81
	Img3	16.47	13.19	13.49	14.98	13.20		22.77	22.51	23.10	19.84	26.05
	Img4	18.11	16.44	16.45	18.61	16.73		21.85	23.02	24.02	19.29	25.45
	Img5	14.02	17.47	15.05	17.19	16.82		23.60	22.03	22.66	16.95	24.05
	Img6	16.14	17.01	16.23	17.01	18.33		21.28	19.34	24.48	19.98	26.74
4	Img1	18.81	20.28	20.50	21.91	19.84	15	27.18	27.98	25.41	21.12	28.75
	Img2	13.79	15.49	15.66	16.20	18.85		21.80	23.76	24.23	21.72	28.12
	Img3	19.16	14.55	15.33	13.29	14.40		24.75	23.32	23.01	22.22	24.36
	Img4	17.36	19.59	19.70	16.93	18.78		26.18	25.04	26.68	19.55	26.30
	Img5	16.85	16.86	16.04	15.09	17.54		24.95	25.34	24.59	19.13	27.88
	Img6	18.23	19.24	16.56	15.16	21.57		24.19	22.35	26.22	20.94	29.15

Table 7.2 The SSIM value for each algorithm at different threshold levels

k	Im	ABC	SCA	SSO	PSO	TSA	Im	ABC	SCA	SSO	PSO	TSA
2	Img1	0.49	0.63	0.67	0.66	0.67	5	0.76	0.76	0.77	0.83	0.76
	Img2	0.42	0.58	0.59	0.57	0.59		0.75	0.82	0.76	0.83	0.74
	Img3	0.57	0.56	0.57	0.55	0.58		0.70	0.79	0.76	0.68	0.78
	Img4	0.55	0.53	0.56	0.56	0.56		0.72	0.65	0.69	0.64	0.74
	Img5	0.34	0.45	0.54	0.55	0.52		0.70	0.71	0.69	0.71	0.66
	Img6	0.61	0.59	0.64	0.64	0.63		0.74	0.72	0.67	0.73	0.72
3	Img1	0.69	0.74	0.73	0.73	0.63	10	0.85	0.84	0.84	0.78	0.89
	Img2	0.67	0.72	0.71	0.68	0.69		0.87	0.89	0.84	0.71	0.90
	Img3	0.54	0.59	0.59	0.68	0.63		0.76	0.80	0.84	0.76	0.89
	Img4	0.65	0.60	0.62	0.66	0.61		0.75	0.80	0.82	0.68	0.85
	Img5	0.55	0.65	0.63	0.66	0.65		0.85	0.85	0.86	0.62	0.87
	Img6	0.66	0.66	0.65	0.65	0.68		0.74	0.75	0.82	0.69	0.87
4	Img1	0.72	0.73	0.76	0.78	0.80	15	0.89	0.89	0.91	0.83	0.93
	Img2	0.65	0.70	0.71	0.76	0.76		0.89	0.89	0.87	0.86	0.93
	Img3	0.70	0.69	0.64	0.49	0.73		0.87	0.83	0.85	0.83	0.94
	Img4	0.61	0.71	0.67	0.61	0.72		0.85	0.84	0.85	0.69	0.91
	Img5	0.70	0.68	0.67	0.50	0.69		0.88	0.90	0.89	0.71	0.91
	Img6	0.67	0.68	0.67	0.57	0.76		0.83	0.79	0.88	0.78	0.93

Table 7.3 The Fitness values for each algorithm among all the threshold values

k	Im	ABC	SCA	SSO	PSO	TSA	k	ABC	SCA	SSO	PSO	TSA
2	Img1	1624.86	1672.28	1747.24	1775.20	1745.58	5	1892.73	1921.03	1912.68	1974.08	1922.16
	Img2	1555.41	1753.00	1714.62	1775.67	1657.25		1929.31	1933.39	1930.63	1959.09	1914.05
	Img3	1605.68	1583.81	1562.32	1607.89	1606.02		1660.91	1679.68	1663.41	1810.17	1675.27
	Img4	1601.48	1590.19	1600.94	1607.94	1601.98		1657.18	1642.98	1637.51	1717.19	1662.73
	Img5	1466.93	1522.85	1602.92	1607.66	1580.52		1659.78	1661.55	1658.54	1808.33	1656.10
	Img6	1577.25	1586.76	1598.72	1606.12	1600.07		1651.34	1663.65	1656.58	1670.55	1662.62
3	Img1	1773.38	1842.07	1843.37	1883.64	1781.61	10	1991.24	1992.44	1994.94	1192.15	2008.33
	Img2	1773.48	1810.68	1863.45	1881.71	1865.45		1970.88	1984.77	1984.58	1822.22	1990.30
	Img3	1607.16	1614.04	1612.37	1647.83	1614.99		1685.39	1666.98	1689.29	1460.41	1686.17
	Img4	1645.03	1635.11	1635.53	1645.22	1627.94		1677.42	1686.77	1692.58	2200.29	1684.96
	Img5	1612.78	1628.44	1638.35	1642.31	1631.06		1686.14	1688.61	1693.40	2233.55	1695.34
	Img6	1640.10	1620.10	1620.33	1646.54	1607.34		1690.70	1688.26	1682.06	1731.41	1689.59
4	Img1	1880.58	1858.13	1886.79	1923.17	1817.34	15	2014.73	2023.15	1997.14	2261.29	2021.60
	Img2	1878.42	1898.42	1902.48	1937.91	1833.93		2019.66	2016.24	2012.85	2885.26	2015.99
	Img3	1651.26	1646.06	1658.88	1676.97	1657.43		1701.72	1698.51	1696.08	1748.13	1695.99
	Img4	1632.19	1670.34	1638.73	1833.79	1654.44		1699.80	1695.43	1698.28	1846.71	1700.81
	Img5	1663.24	1651.87	1660.45	1723.49	1621.37		1695.33	1697.62	1698.03	1104.11	1697.48
	Img6	1639.47	1644.89	1641.35	1803.87	1650.45		1696.10	1694.70	1700.32	1873.52	1696.27

Table 7.4 The CPU time(s) for each algorithm

k	Im	ABC	SCA	SSO	PSO	TSA	k	ABC	SCA	SSO	PSO	TSA
2	Img1	0.34	0.25	0.22	0.17	0.37	5	0.40	0.19	0.34	0.15	0.34
	Img2	0.31	0.19	0.17	0.14	0.29		0.40	0.19	0.28	0.15	0.24
	Img3	0.45	0.32	0.30	0.29	0.31		0.57	0.39	0.35	0.29	0.35
	Img4	0.57	0.36	0.37	0.30	0.38		0.49	0.41	0.34	0.29	0.36
	Img5	0.45	0.32	0.30	0.28	0.24		0.57	0.40	0.36	0.30	0.35
	Img6	0.44	0.31	0.38	0.32	0.32		0.47	0.39	0.33	0.34	0.35
3	Img1	0.33	0.19	0.21	0.16	0.22	10	0.44	0.32	0.29	0.21	0.47
	Img2	0.31	0.19	0.22	0.14	0.30		0.43	0.29	0.25	0.20	0.35
	Img3	0.52	0.36	0.46	0.31	0.42		0.49	0.39	0.37	0.32	0.36
	Img4	0.53	0.37	0.40	0.31	0.45		0.56	0.39	0.41	0.32	0.36
	Img5	0.52	0.36	0.32	0.29	0.23		0.52	0.43	0.38	0.35	0.36
	Img6	0.51	0.36	0.34	0.30	0.24		0.51	0.44	0.39	0.32	0.36
4	Img1	0.31	0.18	0.23	0.18	0.22	15	0.38	0.33	0.26	0.18	0.33
	Img2	0.34	0.18	0.20	0.16	0.22		0.35	0.33	0.29	0.24	0.35
	Img3	0.59	0.39	0.39	0.34	0.22		0.57	0.44	0.40	0.36	0.37
	Img4	0.58	0.40	0.33	0.29	0.49		0.57	0.49	0.46	0.35	0.38
	Img5	0.54	0.38	0.33	0.29	0.45		0.51	0.44	0.41	0.33	0.37
	Img6	0.56	0.37	0.34	0.31	0.39		0.55	0.49	0.43	0.33	0.41

proposed method is increased with increasing the level of threshold which has the better PSNR at levels 4, 5, 10, 15.

In addition, Table 7.2 shows the SSIM results of the algorithms; it can conclude from these results that the SSO, PSO, and SCA have the better value at the threshold 2, 3, and 5 respectively. Meanwhile, the TSA allocates nearly, the second rank, however, at the remaining thresholds (4,10,15) the proposed TSA method allocates the first rank. The fitness value for each algorithm is given in Table 7.3, which can be concluded that the PSO has the higher fitness value overall the thresholds except at threshold 10. Also, it can be seen that most of the algorithms have nearly the same fitness value.

Table 7.4 illustrates the CPU time(s) for each algorithm along at each threshold, where it can be observed that ABC has the largest time to find the threshold, while, PSO is best algorithm regarding CPU time(s). The TSA algorithm is better than other algorithms at threshold 3, 10, and 15, however, at thresholds 2 and 4 it has higher CPU time(s) than other algorithms.

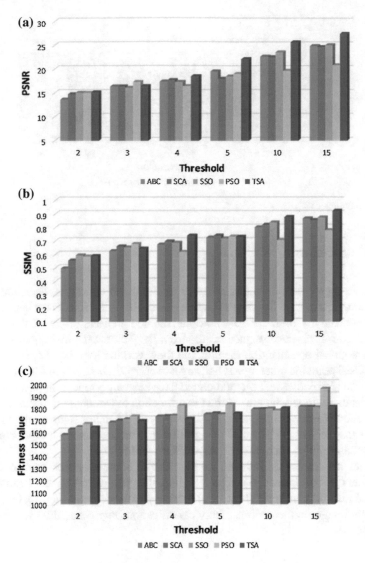

Fig. 7.2 The average of **a** PSNR, **b** SSIM, **c** Fitness value, **d** CPU time(s)

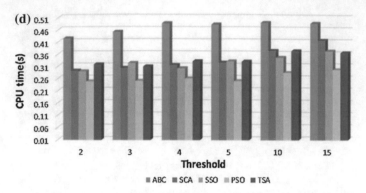

Fig. 7.2 (continued)

7.5 Conclusion

The image segmentation methods based on the combination between the meta-heuristic algorithms and the Otsu's function have more attentions in the last decades. Since they avoid the limitations of traditional method such as stuck at local point. The mean reason for this performance results from the fact that the meta-heuristic algorithms are global optimization method and they working together. So, if one agent stuck at local point the other agents helping it to change it is situation. Based on these concepts an alternative MH method called Tree seed algorithm is applied as image segmentation method which used the Otsu function to evaluate the performance of each solution. To assess the quality of the proposed TSA image segmentation method a set of experimental are executed using different images. Also, it is compared with other four algorithms such as PSO, ABC, SCA, and SSO.

The comparison results indicate the high ability of the proposed TSA to find the multi-level threshold value to produce high quality segmented image. In future the proposed method can applied into different applications including, feature selection, color image segmentation, galaxy classification, improving the Deep learning performance.

References

1. El Aziz MA, Ewees AA, Hassanien AE (2017) Whale optimization algorithm and moth-flame optimization for multilevel thresholding image segmentation. Expert Syst Appl 83:242–256. https://doi.org/10.1016/j.eswa.2017.04.023
2. Bhandari AK, Kumar A, Singh GK (2015) Tsallis entropy based multilevel thresholding for colored satellite image segmentation using evolutionary algorithms. Expert Syst Appl 42:8707–8730. https://doi.org/10.1016/j.eswa.2015.07.025
3. Amerifar S (2015) Iris the picture of health : towards medical diagnosis of diseases based on Iris pattern, pp 120–123

4. Ramakrishnan T, Sankaragomathi B (2017) A professional estimate on the computed tomography brain tumor images using SVM-SMO for classification and MRG-GWO for segmentation. Pattern Recognit Lett 94:163–171. https://doi.org/10.1016/j.patrec.2017.03.026

5. Chelva MS, Samal AK (2017) A comprehensive study of edge detection techniques in image processing applications using particle swarm optimization algorithm. 14:220–228

6. Oliver A, Mu X, Batlle J (2006) Improving clustering algorithms for image segmentation using contour and region information, p 1–6

7. Qi C (2014) Maximum entropy for image segmentation based on an adaptive particle swarm optimization. Appl Math Inf Sci 8:3129–3135. https://doi.org/10.12785/amis/080654

8. Khan W (2013) Image segmentation techniques a survey. J Image Graph 1:166–170. https://doi.org/10.12720/joig.1.4.166-170

9. Alsmadi M (2014) A hybrid firefly algorithm with fuzzy-c mean algorithm for MRI brain segmentation. Am J Appl Sci 11:1676–1691. https://doi.org/10.3844/ajassp.2014.1676.1691

10. Li L, Li D (2008) Fuzzy entropy image segmentation based on particle swarm optimization. Prog Nat Sci 18:1167–1172. https://doi.org/10.1016/j.pnsc.2008.03.020

11. Horng M-H (2010) A multilevel image thresholding using the honey bee mating optimization. Appl Math Comput 215:3302–3310

12. Akay B (2013) A study on particle swarm optimization and artificial bee colony algorithms for multilevel thresholding. Appl Soft Comput J 13:3066–3091. https://doi.org/10.1016/j.asoc.2012.03.072

13. Mlakar U, Potočnik B, Brest J (2016) A hybrid differential evolution for optimal multilevel image thresholding. Expert Syst Appl 65:221–232. https://doi.org/10.1016/j.eswa.2016.08.046

14. Pare S, Kumar A, Bajaj V, Singh GK (2016) A multilevel color image segmentation technique based on cuckoo search algorithm and energy curve. Appl Soft Comput J 47:76–102. https://doi.org/10.1016/j.asoc.2016.05.040

15. Oliva D, Cuevas E, Pajares G et al (2013) Multilevel thresholding segmentation based on harmony search optimization. J Appl Math 2013:24. https://doi.org/10.1155/2013/575414

16. Otsu N (1979) A threshold selection method from gray-level histograms. IEEE Trans Syst Man Cybern 9:62–66. https://doi.org/10.1109/TSMC.1979.4310076

17. Kiran MS (2015) TSA: Tree-seed algorithm for continuous optimization. Expert Syst Appl 42:6686–6698

18. Kennedy J, Eberhart R (1995) Particle swarm optimization. IEEE Int. Conf. Part. swarm Optim. 4:1942–1948

19. Abd El Aziz M, Selim IM, Xiong S (2017) Automatic detection of galaxy type from datasets of galaxies image based on image retrieval approach. Sci Rep 7:1–21. https://doi.org/10.1038/s41598-017-04605-9

20. Mirjalili S (2016) SCA: a sine cosine algorithm for solving optimization problems. Knowl-Based Syst 96:120–133. https://doi.org/10.1016/j.knosys.2015.12.022

21. Abd Elaziz M, Oliva D, Xiong S (2017) An improved opposition-based sine cosine algorithm for global optimization. Expert Syst Appl 90:484–500. https://doi.org/10.1016/j.eswa.2017.07.043

22. Yu JJQ, Li VOK (2015) A social spider algorithm for global optimization. Appl Soft Comput 30:614–627. https://doi.org/10.1016/j.asoc.2015.02.014

23. Abd El Aziz M, Hassanien AE (2017) An improved social spider optimization algorithm based on rough sets for solving minimum number attribute reduction problem. Neural Comput Appl 30:2441–2452. https://doi.org/10.1007/s00521-016-2804-8

24. Wang Z, Bovik AC, Sheikh HR, Simoncelli EP (2004) Image quality assessment: from error visibility to structural similarity. IEEE Trans Image Process 13:600–612. https://doi.org/10.1109/TIP.2003.819861

25. Agrawal S, Panda R, Bhuyan S, Panigrahi BK (2013) Tsallis entropy based optimal multilevel thresholding using cuckoo search algorithm. Swarm Evol Comput 11:16–30. https://doi.org/10.1016/j.swevo.2013.02.001

Chapter 8
Image Segmentation Using Kapur's Entropy and a Hybrid Optimization Algorithm

8.1 Introduction

As was previously mentioned, image segmentation is considered one of the most popular pre-processing technique in image processing. This task attracted many researchers in different applications such as pattern recognition [1, 2], face recognition [3], agriculture [4, 5], satellite image [6] and others [7].

The image segmentation aims to separate the image into its classes that have the same properties (i.e., contrast, color, and others). Therefore, there are several methods have been presented to segment the image. However, the thresholding approach is the most common image segmentation used since they are easiest to implement and more accurate than other methods such as edge detection [8], and clustering algorithms [9]. These methods split pixels of the image into different classes based on the histogram of the given image. Where the pixels that have values lower than the first threshold they belong to the first class, and those pixels have a value lower than the second threshold and greater than the first threshold, belong to the second class and so on [10, 11].

In general, the threshold methods are categorized into two groups; the first group is called bi-level thresholding which separates the image into foreground and background. However, most of the images contain more than two classes, so, the bi-level group is not suitable for this kind of images. To avoid the limitations of the bi-level, the Multi-level thresholding (MLT) methods are used [11, 12]. These MLT methods separate the image into different classes based on more than two thresholds values. This can achieve by using some measures such as the Otsu's method that calculate the intraclass variance to select the best thresholds [13]. Another important technique is the Kapur's entropy [14], in which the aim is to use the thresholds that maximize the entropy of the classes. However, these methods suffer from some limitations, for example; they are time-consuming, so to avoid this limitation the meta-heuristic (MH) methods are used.

© Springer Nature Switzerland AG 2019
D. Oliva et al., *Metaheuristic Algorithms for Image Segmentation:
Theory and Applications*, Studies in Computational Intelligence 825,
https://doi.org/10.1007/978-3-030-12931-6_8

The MH algorithms are simulating the natural process such as biology, physics rule, and other behaviors and they are used to enhance the solutions of different problems. For example, Firefly Algorithm (FA) [15], the Genetic Algorithm (GA) [16], Ant Colony Optimization (ACO) [17], Particle Swarm Optimization (PSO) [18], and Social Spider Optimization (SSO) [19, 20].

Most of these MH methods have been used to find the threshold values to segment the images. In which most of them used the Otsu's or Kapur's methods as fitness function which evaluate the solutions. For example, Artificial Bee Colony (ABC) [10, 21] Honey Bee Mating Optimization (HBMO) [22], Bacterial Foraging Optimization (BFO) [23], SSO [24], Wind Driven Optimization (WDO) [25], Cuckoo Search (CS) [26], and Harmony Search (HS) algorithm [27].

However, most of the previous MH methods still have some drawbacks that affect their quality, generating low accuracy on the results. To overcome these problems, the hybridization between MH algorithms can be used which use different operators to generate better results. Based on these concepts, there are several hybrid MH algorithms have been proposed for image segmentation, such as the PSO is combined with ABC to improve its behavior as in [28]. The hybrid between FA and SSO has been applied in [29]. Also, the hybridization between the DE and the CS has been proposed in [11] in which the DE consider the CS as local search. Also, the ABC algorithm is used the pattern search as local search and used this algorithm for global optimization [30].

Most of the previous hybridization approaches have different combinations such as they can consist of two layers where the first layer contains one MH method to find a good solution and then pass these solutions to the second layer that contains another MH algorithm to update the current solutions. Another approach to combine the operators is to switch between them based on some measures such as probability functions. However, in some situations, these techniques are time-consuming. Thus, this chapter presents an alternative hybrid approach for MLT image segmentation by combining the salp optimization method with the artificial bee colony. Where the SSA has been applied in several applications such as feature selection method, optimization tasks, PID-Fuzzy control design for seismic exited [31], Allocation and Capacity of Renewable Distributed Generators on Distribution Grids [32]. Also, the ABC used in Static Economic Dispatch (SED) problem [33].

In addition, the proposed method (SSAABC) combines the properties of the SSA and ABC. In which the proposed SSAABC begins by generating random solutions that represent the threshold values of a segmented image. Then the objective function for each solution is calculated using the Kapur function, and the best solution is determined. After that, the probability of each solution is computed to determine if the current solution will update using either the SSA or ABC. Where the SSA is applied if the probability is greater than 0.5 otherwise the ABC is used. The process of updating the solution is repeated until the stop conditions are met.

The rest of this chapter is organized as the following: In Sect. 8.2 presents the Background about the image segmentation using threshold technique, Salp swarm algorithm, and Artificial Bee Colony. Section 8.3 introduces the proposed image segmentation

8.2 Background

This section introduces the main concepts for image segmentation and explains the theory of the selected algorithms to optimize the objective function. In this context, there is explained the formulation of the image thresholding problem and then is introduced the entropy of Kapur that is one of the most important techniques for select the best thresholds for image segmentation. After that, the reader will find the basic concepts about SSA and ABC.

8.2.1 Problem Formulation

The basic multi-level thresholding image segmentation can be defined as the process of separate images into its classes. Let there exist $K + 1$ classes un the given image I, so the major process is to find the a K thresholds to determine the classes of I. The mathematical formula of the MLT is given as:

$$C_0 = \{I_{ij} \in I | 0 \le I_{ij} \le t_1 - 1\}$$
$$C_1 = \{I_{ij} \in I | t_1 \le I_{ij} \le t_2 - 1\}$$
$$\ldots\ldots\ldots$$
$$C_K = \{I_{ij} \in I | t_K \le I_{ij} \le L - 1\}$$

(8.1)

where t_k represents the threshold values and it belongs to the interval $[1, K]$ and the L total number of the gray levels. C_k represents the k-th class produced from t_k.

The problem of finding the threshold values is formulated as an optimization problem that aims to maximize/minimize an objective function that defined as:

$$t_1^*, t_2^*, \ldots, t_K^* = \max_{t_1, t_2, \ldots, t_K} F(t_1, t_2, \ldots, t_K)$$

(8.2)

where F is the Kapur's function [34] that defined as:

$$F_{Kap} = \sum_{i=0}^{K} - \sum_{j=t_i}^{t_{i+1}-1} \frac{P_j}{A_j} \ln\left(\frac{P_j}{A_j}\right), P_j = \sum_{j=t_i}^{t_{i+1}-1} P_j$$

$$A_j = \sum_{j=t_i}^{t_{i+1}-1} P_j, P_i = \frac{Fr_i}{N_p}, t_0 = 0, t_{K+1} = L$$

(8.3)

where P_i and Fr_i represents the frequency, and the probability of the i-th gray level of I, respectively. The μ_1 and N_p represent the mean intensity and the total number of pixels of I, respectively.

8.2.2   Artificial Bee Colony

The Artificial Bee Colony (ABC) is a meta-heuristic algorithm that simulates the behavior of honey bees' colony [35]. The ABC contains three types of groups (1) employed bees which searching about the position of the new food sources. (2) onlooker bees which use the information transferred from the first group to select the food source. (3) the group of scout bees which search about the food source in a random way.

Similar to other MH methods, the ABC starts by generating a random a set of N individuals X which represent the solutions of the given problem. These solutions X are called employed bees and they are used to generate a new set of solutions Y using the following equation [35]:

$$y_i = x_i + \alpha_i(x_i - x_k), k = int(r_2 \times N), j = 1, 2, \ldots, Dim \qquad (8.4)$$

Where $\alpha_i \in [-1, 1]$ is a random number, x_k represents the neighbor employed bee of the ith bee. The fitness function $f(x_i)$ and $f(y_i)$ of each x_i and y_i is computed and if the $f(y_i)$ is better than the $f(x_i)$ then y_i is replaced x_i otherwise, the x_i saved.

The onlooker bees received the $f(x_i)$ from the employed bees, and the x_i that has the higher probability of objective function P_i is selected by using the roulette wheel selection method. This probability is computed as [36]:

$$P_i = \frac{fit_i}{\sum_{i=1}^{N} fit_i}, \quad fit_i = \begin{cases} \frac{1}{1+f_i} & f_i > 0 \\ 1 + abs(f_i) & otherwise \end{cases} \qquad (8.5)$$

The onlooker bees using the same method which used by employed bees to update the position of their individuals. The fitness function of each onlooker is calculated, and the fitness functions of the updated solution and the old solution are compared, and the worst one is removed from memory.

Thereafter, the solutions that not enhanced during limitated number of iterations are considered as an abandoned solution which belongs to the scout bees and these solutions are updated as [35]:

$$x_i = x_i^{min} + \left(x_i^{max} - x_i^{min}\right) * r_2, \qquad (8.6)$$

where x_j^{min} and x_j^{max} represent lower and the upper boundaries for x_i respectively, and r_2 represents a random number. This updated solution x_i becomes employed bee and used in the next iterations.

8.2.3 Salp Swarm Algorithm

The Salp Swarm Algorithm (SSA) is a meta-heuristic algorithm which simulates the behavior of a family of Salps [37]. The SSA has two groups called the leader and the followers where the salp in the front of the chain is called the leader, while others represent the followers.

The salp search for the food source which represents the target. In general, the position of leader at the j th dimension (x_j^1 is updating using the following equation [37]:

$$x_j^1 = \begin{cases} F_j + c_1(U_j - L_j) \times c_2 + L_j & if \ c_2 \le 0 \\ F_j - c_1(U_j - L_j) \times c_2 + L_j & if \ c_2 >; 0 \end{cases} \tag{8.7}$$

Where F_j, L_j and U_j represent the target, lower boundary and upper boundary at jth dimension, respectively. while, the c_2 and c_3 represent a random number which used to maintain the search domain. The c_1 is used to balance between the exploitation phase and the exploration phase and it can be updated during each iterations t until reached the maximum number of iterations t_{max} as:

$$c_1 = 2e^{-\left(\frac{4t}{t_{max}}\right)^2} \tag{8.8}$$

After updating the leader's position, the SSA starts to update the position of the follower x_j^i as:

$$x_j^i = \frac{1}{2}\left(x_j^i - x_j^{i-1}\right), \quad i > 1 \tag{8.9}$$

8.3 Proposed Approach

The proposed image segmentation method based on the combination between the Salp swarm algorithm and artificial bee colony is presented in this section. The proposed method, called SSAABC, used the Kapur function as the objective function to evaluate the performance of each solution.

The input to the proposed SSAABC method is the tested image (I), and the SSAABC compute the histogram of I. Whereas the global solution F is the output of the proposed method which represents the threshold values. The next step is to construct a set of solutions in the interval $[I_{min}, I_{max}]$. where I_{min} and I_{max} is the boundaries of search space as in the following equation:

$$x_i = I_{min} + rand(I_{max} - I_{min}) \tag{8.10}$$

Where *rand* represents a random number in the interval [0, 1]. The next step is to use the kapur function to compute the quality of each solution x_i and find the best solution. Thereafter, the probability of each solution is computed

$$prob_i = \frac{Fit_i}{\sum_{i=1}^{N} Fit_i} \tag{8.11}$$

where Fit_i represents the objective function of the i th solution. If the $prob_i \geq 0.5$ this mean the SSA is used to update the current solution; otherwise the ABC is used to update the solution.

After update all the solutions the best solution is determined which has a higher objective value. The process of updating the population based on the probability is performed again until the termination criterion is met. The final steps of the proposed approach are given in Figure.

8.4 Experimental and Results

The performance of the proposed method to improve the image segmentation by finding the optimal thresholds is evaluated in this section. Where a set of images are used, also, the results of the proposed method is compared with other algorithms including, SSO, SCA, ABC, and PSO.

8.4.1 Benchmark Image

In this study, a set of the six images from Berkeley University database are used. These images have different characteristics, for example, irregular distributions that can be observed Fig. 8.1.

8.4.2 Parameter Settings

The performance of the proposed algorithm is compared with three meta-heuristic algorithms called PSO, ABC, and SCA. The parameters of each method are assigned as an original reference. Meanwhile, the other parameters such as population size are 20, and dimension is set as 2, 4, 10 and 15. Also, the maximum number of iterations is set to 100. The implementation of all algorithms is performed over MATLAB 2017b on windows 10 (64 bit).

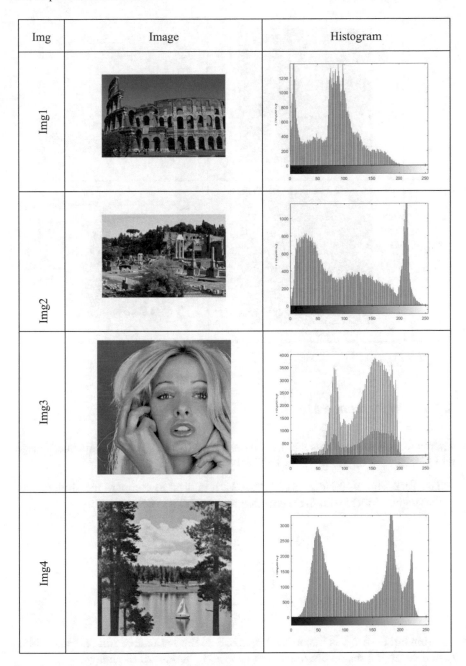

Fig. 8.1 Selected images for graphical analysis

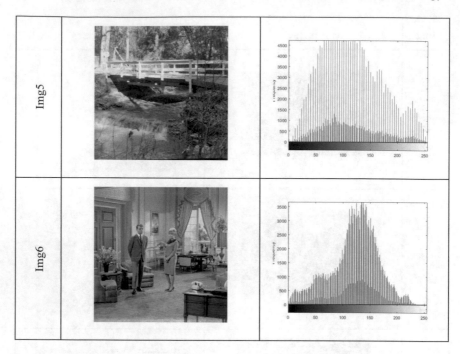

Fig. 8.1 (continued)

8.4.3   Performance Metrics

There are a set of measures are used to evaluate the quality of the segmented image which including objective value, PSNR, SSIM, and CPU time(s):

– The Peak-Signal-to-Noise Ratio (PSNR): it is used to compare the similarity of the original image with the segmented image, and it is defined as:

$$PSNR = 20 \log_{10}\left(\frac{255}{RMSE}\right), \text{(dB)}$$

$$RMSE = \sqrt{\frac{\sum_{i=1}^{ro}\sum_{j=1}^{co}(I_{Gr}(i,j) - I_s(i,j))}{ro \times co}} \tag{8.12}$$

From Eq. 8.12 I_{Gr} is the original image, I_s is the segmented image. Meanwhile, *ro* and *co* are the maximum number of rows and columns of the image.

– The Structure Similarity Index (SSIM) is used to compare the structures of the
– original segmented image with the segmented image [38], and it is defined as:

$$SSIM(I_{or}, I_{th}) = \frac{\left(2\mu_{I_{or}}\mu_{I_{th}} + C1\right)\left(2\sigma_{I_{or}I_{th}} + C2\right)}{\left(\mu_{I_{or}}^2 + \mu_{I_{th}}^2 + C1\right)\left(\sigma_{I_{or}}^2 + \sigma_{I_{th}}^2 + C2\right)}$$

$$\sigma_{I_{th}I_{or}} = \frac{1}{N-1}\sum_{i=1}^{N}\left(I_{or_i} + \mu_{I_{or}}\right)\left(I_{th_i} + \mu_{I_{th}}\right) \qquad (8.13)$$

where the mean of the original image is $\mu_{I_{or}}$ and the mean of the thresholded image is represented by $\mu_{I_{th}}$. In the same way, for each image, the values of $\sigma_{I_{Gr}}$ and $\sigma_{I_{th}}$ correspond to the standard deviation. $C1$ and $C2$ are constants used to avoid instability when $\mu_{I_{Gr}}^2 + \mu_{I_{th}}^2 \approx 0$. The values of $C1$ and $C2$ are set to 0.065 considering the experiments of [26].

8.4.3.1   Results and Discussion

The results of the proposed method and the other methods are given in different Tables and Figures. In which Table 8.1 and Fig. 8.2 show the results of each algorithm in terms of PSNR value. From these results, it can be concluded that the proposed ABCSSA algorithm achieves the first rank regarding the PSNR with 20 cases from 36 cases. While, ABC, SCA, and PSO achieve the second, third and fourth rank

Table 8.1 The PSNR value of each algorithm

k	Img	ABC	SCA	SSO	ABC-SSA	PSO	k	Img	ABC	SCA	SSO	ABC-SSA	PSO
2	Img1	10.80	10.81	11.11	17.37	10.84	5	Img1	16.65	15.01	20.88	21.33	21.42
	Img2	13.79	11.58	11.62	13.42	11.92		Img2	17.16	16.62	18.41	18.33	16.89
	Img3	12.20	11.41	11.29	13.49	11.71		Img3	18.03	14.70	17.35	19.09	18.31
	Img4	11.54	11.79	13.38	15.83	10.80		Img4	18.20	19.78	19.06	19.85	16.16
	Img5	11.82	12.53	12.19	14.29	11.31		Img5	18.02	17.01	15.87	18.35	18.69
	Img6	12.39	12.30	12.23	15.38	12.29		Img6	20.59	18.99	17.08	20.09	15.60
3	Img1	17.15	15.48	18.33	20.41	18.30	10	Img1	19.78	24.60	24.39	25.55	24.70
	Img2	16.24	15.25	14.20	15.62	14.48		Img2	23.44	23.36	23.57	23.20	21.05
	Img3	14.06	16.64	13.09	14.74	14.18		Img3	21.41	20.80	21.84	21.59	22.99
	Img4	15.79	16.65	16.18	18.18	16.38		Img4	22.64	23.50	22.28	23.91	21.07
	Img5	13.56	15.74	10.88	13.68	14.04		Img5	24.88	22.39	23.81	24.40	21.98
	Img6	16.68	16.84	15.68	16.75	15.88		Img6	25.34	18.30	21.47	25.22	25.77
4	Img1	19.21	18.14	18.76	21.31	19.10	15	Img1	23.10	27.32	27.59	26.91	26.12
	Img2	16.02	17.20	16.13	16.75	16.93		Img2	26.93	24.80	23.94	26.76	26.25
	Img3	15.15	18.21	16.55	17.27	15.45		Img3	26.79	23.17	26.51	26.53	25.71
	Img4	18.69	18.44	14.64	16.20	17.94		Img4	24.63	25.95	25.96	27.89	26.41
	Img5	16.53	16.07	13.81	16.11	16.07		Img5	24.49	25.60	23.90	25.13	25.54
	Img6	19.03	15.14	17.32	17.60	17.54		Img6	27.52	26.22	25.13	28.39	24.10

Fig. 8.2 Average of fitness function

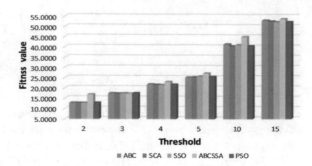

Table 8.2 The results of each algorithm based on the SSIM measure

k	Img	ABC	SCA	SSO	ABC-SSA	PSO	k	Img	ABC	SCA	SSO	ABC-SSA	PSO
2	Img1	0.147	0.149	0.172	0.654	0.150	5	Img1	0.637	0.468	0.764	0.765	0.753
	Img2	0.407	0.387	0.341	0.552	0.391		Img2	0.737	0.582	0.614	0.662	0.547
	Img3	0.369	0.386	0.361	0.573	0.383		Img3	0.575	0.685	0.583	0.638	0.629
	Img4	0.316	0.326	0.409	0.557	0.273		Img4	0.646	0.675	0.655	0.665	0.568
	Img5	0.302	0.349	0.325	0.544	0.268		Img5	0.690	0.623	0.554	0.681	0.680
	Img6	0.428	0.435	0.435	0.631	0.436		Img6	0.737	0.671	0.626	0.739	0.652
3	Img1	0.632	0.560	0.678	0.746	0.672	10	Img1	0.774	0.834	0.831	0.834	0.847
	Img2	0.545	0.506	0.532	0.633	0.558		Img2	0.878	0.838	0.894	0.889	0.817
	Img3	0.543	0.570	0.483	0.651	0.456		Img3	0.732	0.701	0.784	0.816	0.803
	Img4	0.527	0.562	0.545	0.661	0.558		Img4	0.773	0.799	0.772	0.787	0.732
	Img5	0.442	0.571	0.257	0.453	0.494		Img5	0.871	0.819	0.847	0.894	0.831
	Img6	0.608	0.637	0.619	0.650	0.627		Img6	0.814	0.667	0.770	0.877	0.854
4	Img1	0.721	0.670	0.705	0.792	0.685	15	Img1	0.892	0.920	0.905	0.925	0.871
	Img2	0.569	0.584	0.637	0.746	0.572		Img2	0.931	0.929	0.831	0.947	0.915
	Img3	0.515	0.648	0.696	0.658	0.490		Img3	0.877	0.843	0.856	0.876	0.847
	Img4	0.656	0.641	0.508	0.547	0.626		Img4	0.851	0.851	0.842	0.872	0.858
	Img5	0.622	0.595	0.450	0.513	0.598		Img5	0.870	0.888	0.859	0.920	0.896
	Img6	0.676	0.637	0.667	0.659	0.629		Img6	0.892	0.864	0.816	0.902	0.794

followed by the SSO in the last rank. Also, from Fig. 8.1 it can be observed that the proposed ABCSSA approach is better than others except at level 3 where the PSO is the better (Table 8.2).

According to the SSIM values the proposed ABCSSA method has the first rank with 28 cases from 36 cases, while, the ABC achieves the second rank with five cases followed by the SCA with three cases, the same results can be concluded from Fig. 8.2 which shows the average of SSIM along each threshold level (Fig. 8.3).

Also, Table 8.3 illustrates the fitness function for each algorithm from this table it can be observed that from the total number of experiments (36) the proposed achieves the first rank at 28 cases. While both of the ABC and PSO has three cases and the SCA

Fig. 8.3 Average of PSNR

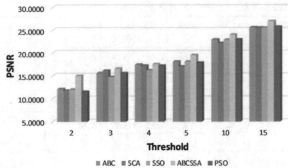

Table 8.3 The fitness function value for each algorithm

k	Img	ABC	SCA	SSO	ABC-SSA	PSO	k	Img	ABC	SCA	SSO	ABC-SSA	PSO
2	Img1	12.79	12.70	12.77	17.41	12.80	5	Img1	24.94	25.38	25.74	26.67	25.72
	Img2	12.74	12.66	12.73	17.22	12.65		Img2	24.73	24.59	26.06	26.93	25.27
	Img3	12.80	12.81	12.54	16.92	12.80		Img3	25.86	25.06	25.84	28.64	25.37
	Img4	12.80	12.79	12.74	16.56	12.69		Img4	24.34	25.86	25.78	28.61	25.79
	Img5	12.80	12.55	12.74	16.66	12.81		Img5	25.67	25.62	24.89	25.49	25.42
	Img6	12.81	12.80	12.78	16.18	12.79		Img6	25.20	24.84	25.73	26.49	25.40
3	Img1	17.36	17.03	17.35	18.23	17.58	10	Img1	40.68	40.95	41.25	43.12	41.97
	Img2	17.27	17.02	17.57	18.90	17.22		Img2	40.39	40.19	41.84	42.61	40.66
	Img3	17.66	17.03	17.47	16.10	17.28		Img3	41.62	40.06	40.36	40.54	40.38
	Img4	17.43	17.36	17.35	16.50	17.67		Img4	41.06	41.49	40.13	45.52	39.28
	Img5	17.20	17.49	17.08	17.21	17.40		Img5	41.74	39.34	41.87	43.39	39.70
	Img6	17.63	17.68	17.49	16.54	17.53		Img6	42.72	40.11	40.85	54.66	40.82
4	Img1	21.60	21.74	21.63	22.58	21.93	15	Img1	52.38	52.31	51.02	52.48	50.95
	Img2	21.64	21.82	21.50	23.63	21.70		Img2	53.82	52.49	51.99	54.81	55.21
	Img3	21.92	21.41	21.57	22.54	21.59		Img3	53.11	53.28	53.68	56.75	52.77
	Img4	21.82	21.47	21.15	23.87	21.62		Img4	52.91	51.58	53.35	54.21	51.45
	Img5	21.55	21.39	21.34	22.55	21.47		Img5	52.03	52.56	52.04	52.68	53.00
	Img6	21.90	21.46	21.61	22.43	21.61		Img6	54.01	53.37	51.76	51.72	50.37

has the highest fitness function at two cases. Moreover, the CPU time(s) required by the proposed method is the smaller among all the comparative algorithms, whereas, the PSO requires more time to reach the optimal solution as given in Table 8.4. The same information can be concluded from Fig. 8.4 which shows the average of the CPU time(s) along each threshold (Fig. 8.5).

Table 8.4 The CPU time(S) for each algorithm

k	Img	ABC	SCA	SSO	ABC-SSA	PSO	k	Img	ABC	SCA	SSO	ABC-SSA	PSO
2	Img1	0.47	0.37	0.22	0.14	0.60	5	Img1	0.42	0.29	0.29	0.18	0.62
	Img2	0.31	0.20	0.20	0.14	0.39		Img2	0.44	0.28	0.31	0.15	0.65
	Img3	0.45	0.36	0.34	0.29	0.52		Img3	0.57	0.48	0.40	0.29	0.83
	Img4	0.54	0.40	0.35	0.30	0.54		Img4	0.50	0.42	0.38	0.31	0.80
	Img5	0.58	0.43	0.39	0.30	0.59		Img5	0.48	0.45	0.41	0.32	0.76
	Img6	0.45	0.41	0.38	0.28	0.52		Img6	0.53	0.46	0.39	0.29	0.80
3	Img1	0.33	0.22	0.24	0.17	0.54	10	Img1	0.46	0.40	0.38	0.17	1.01
	Img2	0.32	0.24	0.26	0.23	0.46		Img2	0.41	0.41	0.38	0.20	0.99
	Img3	0.48	0.37	0.38	0.29	0.61		Img3	0.53	0.50	0.48	0.34	1.15
	Img4	0.55	0.43	0.37	0.29	0.66		Img4	0.54	0.51	0.46	0.33	1.12
	Img5	0.46	0.37	0.35	0.29	0.60		Img5	0.64	0.57	0.52	0.33	1.12
	Img6	0.48	0.46	0.45	0.29	0.62		Img6	0.55	0.52	0.46	0.34	1.10
4	Img1	0.32	0.26	0.22	0.20	0.55	15	Img1	0.44	0.46	0.37	0.19	1.37
	Img2	0.43	0.32	0.26	0.18	0.59		Img2	0.41	0.43	0.54	0.19	1.33
	Img3	0.47	0.38	0.44	0.34	0.73		Img3	0.63	0.62	0.55	0.35	1.47
	Img4	0.47	0.44	0.37	0.31	0.73		Img4	0.72	0.67	0.57	0.39	1.46
	Img5	0.49	0.41	0.39	0.29	0.67		Img5	0.63	0.67	0.57	0.37	1.53
	Img6	0.55	0.44	0.44	0.32	0.75		Img6	0.63	0.67	0.61	0.38	1.56

Fig. 8.4 Average of SSIM

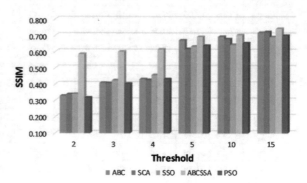

8.5   Conclusion

In this chapter, an alternative image segmentation image segmentation is proposed. The proposed method depends on improvement of the performance of the salp swarm algorithm using the artificial bee colony. In general, the ABC aims to enhance the solutions that have a worst objective function value that is determined by computing the probability of the objective function where the Kapur's function is considered as an objective function. The experiment results illustrated the performance of the

Fig. 8.5 Average of CPU time(s)

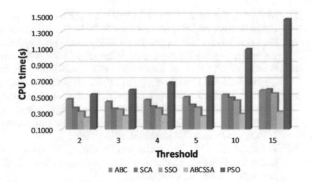

proposed SSAABC method to segmented six images is better than the other four algorithms namely SCA, SSO, ABC, and PSO.

Based on the previous discussion of the quality of the proposed SSAABC, it can be applied in future works to different applications including feature selection, solar cell parameter estimation, face recognition and many other applications.

References

1. Kashiha MA, Bahr C, Ott S et al (2013) Automatic identification of marked pigs in a pen using image pattern recognition. Lect Notes Comput Sci (including Subser Lect Notes Artif Intell Lect Notes Bioinformatics) 7887 LNCS:205–212. https://doi.org/10.1007/978-3-642-38628-2_24
2. Kaur D, Kaur Y (2014) Intelligent medical image segmentation using FCM, GA and PSO. Int J Comput Sci Inf Technol 5:6089–6093
3. Shah JH, Sharif M, Raza M, Azeem A (2013) A survey: linear and nonlinear PCA based face recognition techniques. Int Arab J Inf Technol 10:536–545
4. Ðokić A, Jović S (2017) Evaluation of agriculture and industry effect on economic health by ANFIS approach. Phys A Stat Mech its Appl 479:396–399. https://doi.org/10.1016/j.physa.2017.03.022
5. Campos Y, Sossa H, Pajares G (2016) fields with plants, soil and object discrimination. Precis Agric. https://doi.org/10.1007/s11119-016-9483-4
6. Sarkar S, Das S, Chaudhuri SS (2016) Hyper-spectral image segmentation using Rényi entropy based multi-level thresholding aided with differential evolution. Expert Syst Appl 50:120–129. https://doi.org/10.1016/j.eswa.2015.11.016
7. Dhanya GS, Raj RJS, Sivasamy P (2017) Image segmentation using swarm intelligence based graph partitioning, pp 17–22
8. Gonzalez CI, Melin P, Castro JR et al (2016) Optimization of interval type-2 fuzzy systems for image edge detection. Appl Soft Comput J 47:631–643. https://doi.org/10.1016/j.asoc.2014.12.010
9. Oliver A, Mu X, Batlle J (2006) Improving clustering algorithms for image segmentation using contour and region information, pp 1–6
10. Kumar A, Kumar V, Kumar A, Kumar G (2014) Expert systems with applications Cuckoo search algorithm and wind driven optimization based study of satellite image segmentation for multilevel thresholding using Kapur's entropy. Expert Syst Appl 41:3538–3560. https://doi.org/10.1016/j.eswa.2013.10.059

11. Mlakar U, Potočnik B, Brest J (2016) A hybrid differential evolution for optimal multilevel image thresholding. Expert Syst Appl 65:221–232. https://doi.org/10.1016/j.eswa.2016.08.046

12. Zhang Y, Wu L (2011) Optimal multi-level thresholding based on maximum Tsallis entropy via an artificial bee colony approach. Entropy 13:841–859. https://doi.org/10.3390/e13040841

13. Otsu N (1979) A threshold selection method from gray-level histograms. IEEE Trans Syst Man Cybern 9:62–66. https://doi.org/10.1109/TSMC.1979.4310076

14. Kapur JN, Sahoo PK, Wong AK (1985) A new method for gray-level picture thresholding using the entropy of the histogram. Comput Vision Graph Image Process 29:273–285

15. Yang XS, He X (2013) Firefly algorithm: recent advances and applications. Int J Swarm Intell 1:36. https://doi.org/10.1504/IJSI.2013.055801

16. Ramezani F, Lu J, Taheri J, Hussain FK (2015) Evolutionary algorithm-based multi-objective task scheduling optimization model in cloud environments. World Wide Web 18:1–23. https://doi.org/10.1007/s11280-015-0335-3

17. Wang X-N, Feng Y-J, Feng Z-R (2005) Ant colony optimization for image segmentation. In: Proceedings of international conference on machine learning and cybernetics, vol 9, pp 5355–5360. https://doi.org/10.1109/icmlc.2005.1527890

18. Kennedy J, Eberhart RC (1995) Particle swarm optimization. In: 1995 Proceedings of IEEE International Conference on Neural Networks, vol 4, pp 1942–1948. https://doi.org/10.1109/icnn.1995.488968

19. El Aziz MA, Hassanien AE (2016) An improved social spider optimization algorithm based on rough sets for solving minimum number attribute reduction problem. Neural Comput Appl

20. El Aziz MA, Oliva D (2017) An opposition-based social-spider optimization for feature selection, pp 1–22

21. Akay B (2013) A study on particle swarm optimization and artificial bee colony algorithms for multilevel thresholding. Appl Soft Comput J 13:3066–3091. https://doi.org/10.1016/j.asoc.2012.03.072

22. Horng M-H (2010) A multilevel image thresholding using the honey bee mating optimization. Appl Math Comput 215:3302–3310

23. Bakhshali MA, Shamsi M (2014) Segmentation of color lip images by optimal thresholding using bacterial foraging optimization (BFO). J Comput Sci 5:251–257

24. Agarwal P, Singh R, Kumar S, Bhattacharya M (2016) Social spider algorithm employed multilevel thresholding segmentation approach. In: Proceedings of first international conference on information and communication

25. Bayraktar Z, Komurcu M, Bossard JA, Werner DH (2013) The wind driven optimization technique and its application in electromagnetics. IEEE Trans Antennas Propag 61:2745–2757

26. Agrawal S, Panda R, Bhuyan S, Panigrahi BK (2013) Tsallis entropy based optimal multilevel thresholding using cuckoo search algorithm. Swarm Evol Comput 11:16–30. https://doi.org/10.1016/j.swevo.2013.02.001

27. Oliva D, Cuevas E, Pajares G et al (2013) Multilevel thresholding segmentation based on harmony search optimization. J Appl Math https://doi.org/10.1155/2013/575414

28. Liu Y, Hu K, Zhu Y, Chen H (2014) A novel method for image segmentation based on nature inspired algorithm. In: Huang DS, Han K, Gromiha M (eds) Intelligent Computing in Bioinformatics ICIC 2014, vol 8590. Lecture Notes in Computer Science. Springer, Cham

29. Mohamed AE, Ahmed AE, Aboul Ella H (2016) Hybrid Swarms Optimization Based Image Segmentation. Springer International Publishing

30. Kang F, Li J, Li H (2013) Artificial bee colony algorithm and pattern search hybridized for global optimization. Appl Soft Comput J 13:1781–1791. https://doi.org/10.1016/j.asoc.2012.12.025

31. Mahdi S, Baygi H, Karsaz A (2018) A hybrid optimal PID-Fuzzy control design for seismic exited structural system against earthquake: a Salp swarm algorithm, pp 220–225

32. Tolba M, Rezk H, Diab AAZ (2018) A novel robust methodology based Salp swarm algorithm for allocation and capacity of renewable distributed generators on distribution grids. Energies 11:2556. https://doi.org/10.3390/en11102556

33. Hemamalini S, Simon SP (2011) Dynamic economic dispatch using artificial bee colony algorithm for units with valve-point effect: 70–81. https://doi.org/10.1002/etep
34. Kapur JN, Sahoo PK, Wong AKC (1985) A new method for gray-level picture thresholding using the entropy of the histogram. Comput Vision Graph Image Process 29:273–285
35. Karaboga D, Gorkemli B, Ozturk C, Karaboga N (2014) A comprehensive survey: Artificial Bee Colony (ABC) algorithm and applications. Artif Intell Rev 42:21–57. https://doi.org/10.1007/s10462-012-9328-0
36. Dieleman S, Willett KW, Dambre J (2015) Rotation-invariant convolutional neural networks for galaxy morphology prediction. Mon Not R Astron Soc 450:1441–1459. https://doi.org/10.1093/mnras/stv632
37. Mirjalili S, Gandomi AH, Zahra S, Saremi S (2017) Salp Swarm Algorithm: a bio-inspired optimizer for engineering design problems. Adv Eng Softw 0:1–29. https://doi.org/10.1016/j.advengsoft.2017.07.002
38. Wang Z, Bovik AC, Sheikh HR, Simoncelli EP (2004) Image quality assessment: from error visibility to structural similarity. IEEE Trans Image Process 13:600–612. https://doi.org/10.1109/TIP.2003.819861

Chapter 9
Tsallis Entropy for Image Thresholding

9.1 Introduction

As was previously mentioned, segmentation is one of the basic steps of an image analysis system, and consists in separating objects from each other, by considering characteristics contained in a digital image. In general, thresholding is one of the easiest methods used for segmentation. There are two thresholding approaches, namely bi-level and multilevel. In bi-level thresholding (BT), it is only needed a threshold value to separate the two objects of an image (e.g. foreground and background). For bi-level thresholding there exist two classical methods: the first one, proposed by Otsu [1], maximizes the variance between classes, whereas the second one, proposed by Kapur [2], uses the entropy maximization to measure the homogeneity among classes. Their efficiency and accuracy have been already proved by segmenting pixels into two classes [3]. Since, BT is not enough for complex images with different objects, both methods, Otsu and Kapur, can be expanded for multilevel thresholding. However, their computational complexity is increased, and also its accuracy decreases with each new threshold added into the searching process [3, 4].

The Tsallis entropy (TE), proposed in [5], is known as the non-extensive entropy, and can be considered as an extension of Shannon's entropy. Recently, there exist several studies that report similarities among the Tsallis, the Shannon and the Boltzmann-Gibbs entropies [4, 6–8]. Different to the Otsu and Kapur methods, the Tsallis entropy produces a functional formulation whose accuracy does not depend on the number of threshold points [8]. In the process of image segmentation, under the TE perspective, it is selected a set of threshold values that maximize the TE functional formulation, so that each pixel is assigned to a determined class according to its corresponding threshold points. TE gives excellent results in bi-level thresholding. However, when it is applied to multilevel thresholding (MT), its evaluation becomes computationally expensive, since each threshold point adds restrictions, multimodality and complexity to its functional formulation. Therefore, in the process of finding the appropriate threshold values, it is desired to limit the number of evaluations of the TE objec-

© Springer Nature Switzerland AG 2019
D. Oliva et al., *Metaheuristic Algorithms for Image Segmentation: Theory and Applications*, Studies in Computational Intelligence 825, https://doi.org/10.1007/978-3-030-12931-6_9

tive function. Under such circumstances, most of the optimization algorithms do not seem to be suited to face such problems as they usually require many evaluations before delivering an acceptable result.

This chapter is presents a multilevel thresholding method that uses the Electromagnetism-Like Algorithm (EMO) to find the best threshold values. EMO is a population-based evolutionary method which was firstly introduced by Birbil and Fang [9] to solve unconstrained optimization problems. The algorithm emulates the attraction–repulsion mechanism between charged particles within an electro-magnetism field. Each particle represents a solution and carries a certain amount of charge which is proportional to its fitness value. In turn, solutions are defined by position vectors which give real positions for particles within a multi-dimensional space. Moreover, objective function values of particles are calculated considering such position vectors. Each particle exerts repulsion or attraction forces over other members in the population; the resultant force acting over a particle is used to update its position. Clearly, the idea behind the EMO methodology is to move particles towards the optimum solution by exerting attraction or repulsion forces among them. Different to other evolutionary methods, EMO exhibits interesting search capabilities such as fast convergence still keeping its ability to avoid local minima in high modality environments [10–12]. Some interesting studies [13–16] demonstrate that the EMO algorithm presents the best balance between optimization results and demand of function evaluations. Such characteristics have attracted the attention of the evolutionary computation community, so that it has been effectively applied to solve a wide range of engineering problems such as flow-shop scheduling [17], communications [18], vehicle routing [19], array pattern optimization in circuits [20], neural network training [21], image processing and control systems [22].

In the presented approach, the EMO algorithm is used to find the optimal threshold values by maximizing the Tsallis entropy. As a result, the proposed algorithm can substantially reduce the number of function evaluations preserving the good search capabilities of an evolutionary method. In our approach, the algorithm uses as particles the encoding of a set of candidate threshold points. The TE objective function evaluates the segmentation quality of the candidate threshold points. Guided by the values of this objective function, the set of encoded candidate solutions are modified by using the EMO operators so that they can improve their segmentation quality as the optimization process evolves. In comparison to other similar algorithms, the proposed method deploys better segmentation results yet consuming less TE function evaluations.

The rest of the chapter is organized as follows. In Sect. 9.2, the standard EMO algorithm is introduced. Section 9.3 gives a simple description of the Tsallis entropy method. Section 9.4 explains the implementation of the proposed algorithm. Section 9.5 discusses experimental results and comparisons after testing the proposal over a set of benchmark images. Finally, in Sect. 9.6 the conclusions are discussed.

9.2 Electromagnetism—Like Optimization Algorithm

The Electromagnetism-Like Optimization (EMO) is a population-based evolutionary method which was firstly introduced by Birbil and Fang [9] to solve unconstrained optimization problems. Different to other evolutionary algorithms, EMO exhibits interesting search capabilities such as fast convergence still keeping its ability to avoid local minima in high modality environments [10–12]. Different studies [13, 14, 16, 23] demonstrate that the EMO algorithm presents the best balance between optimization results and demand of function evaluations. From an implementation point of view, EMO utilizes N different n-dimensional points $x_{i,t}$, $i = 1, 2, \ldots, n$, as a population for searching the feasible set $\mathbf{X} = \{x \in \mathbb{R}^n | l_i \leq x \leq u_i\}$, where t denotes the number of iteration (or generation) of the algorithm. The initial population $\mathbf{Sp}_t = \{x_{1,t}, x_{2,t}, \ldots, x_{N,t}\}$ (being $t = 1$), is taken of uniformly distributed samples of the search region, $\mathbf{X}$. We denote the population set at the t-th iteration by $\mathbf{Sp}_t$, and the members of $\mathbf{Sp}_t$ changes with t. After the initialization of $\mathbf{Sp}_t$, EMO continues its iterative process until a stopping condition (e.g. the maximum number of iterations) is met. An iteration of EMO consists of two main steps: in the first step, each point in $\mathbf{Sp}_t$ moves to a different location by using the attraction-repulsion mechanism of the electromagnetism theory [13]. In the second step, points moved by the electromagnetism principle are further perturbed locally by a local search and then become members of $\mathbf{Sp}_{t+1}$ in the $(t+1)$-th iteration. Both the attraction-repulsion mechanism and the local search in EMO are responsible for driving the members, $x_{i,t}$, of $\mathbf{Sp}_t$ to the close proximity of the global optimum.

As with the electromagnetism theory for charged particles, each point $x_{i,t} \in \mathbf{Sp}_t$ in the search space $\mathbf{X}$ is assumed as a charged particle where the charge of a point is computed based on its objective function value. Points with better objective function value have higher charges than other points. The attraction-repulsion mechanism is a process in EMO by which points with more charge attract other points from $\mathbf{Sp}_t$, and points with less charge repel other points. Finally, a total force vector F_i^t, exerted on a point (e.g. the i-th point $x_{i,t}$) is calculated by adding these attraction—repulsion forces, and each $x_{i,t} \in \mathbf{Sp}_t$ is moved in the direction of its total force to the location $y_{i,t}$. A local search is used to explore the vicinity of the each particle according to its fitness. The members, $x_{i,t+1} \in \mathbf{Sp}_{t+1}$, of the $(t + 1)$-th iteration are then found by using:

$$x_{i,t+1} = \begin{cases} y_{i,t} & \text{if } f(y_{i,t}) \leq f(z_{i,t}) \\ z_{i,t} & \text{otherwise} \end{cases} \tag{9.1}$$

Algorithm 9.1 shows the general scheme of EMO. We also provided the description of each step following the algorithm.

Algorithm 9.1 [EMO (N, $Iter_{\max}$, $Iter_{local}$, δ)]

1. Input parameters: Maximum number of iteration $Iter_{\max}$, values for the local search parameter such $Iter_{local}$ and δ, and the size N of the population.
2. Initialize: set the iteration counter $t = 1$, initialize the number of $\mathbf{Sp}_t$ uniformly in $\mathbf{X}$ and identify the best point in $\mathbf{Sp}_t$.
3. while $t < Iter_{\max}$ do
4. $\quad F_i^t \leftarrow \text{CalcF}(\mathbf{Sp}_t)$
5. $\quad y_{i,t} \leftarrow \text{Move}(x_{i,t}, F_i^t)$
6. $\quad z_{i,t} \leftarrow \text{Local}(Iter_{local}, \delta, y_{i,t})$
7. $\quad x_{i,t+1} \leftarrow \text{Select}(\mathbf{Sp}_{t+1}, y_{i,t}, z_{i,t})$
8. end while

Input parameters (Line 1): EMO algorithm runs for $Iter_{\max}$ iterations. In the local search phase, $n \times Iter_{local}$ is the maximum number of locations $z_{i,t}$, within a δ distance of $y_{i,t}$, for each i dimension.

Initialize (Line 2): The points $x_{i,t}$, $t = 1$, are selected uniformly in $\mathbf{X}$, i.e. $x_{i,1} \sim Unif(\mathbf{X})$, $i = 1, 2, \ldots, N$, where $Unif$ represents the uniform distribution. The objective function values $f(x_{i,t})$ are computed, and the best point is identified for minimization:

$$x_t^B = \arg \min_{x_{i,t} \in \mathbf{S}_t} \left\{ f(x_{i,t}) \right\} \qquad (9.2)$$

and for maximization:

$$x_t^B = \arg \max_{x_{i,t} \in \mathbf{S}_t} \left\{ f(x_{i,t}) \right\} \qquad (9.3)$$

Calculate force (Line 4): In this step, a charged-like value ($q_{i,t}$) is assigned to each point ($x_{i,t}$). The charge $q_{i,t}$ of $x_{i,t}$ depends on $f(x_{i,t})$ and points with better objective function have more charge than others. The charges are computed as follows:

$$q_{i,t} = \exp \left(-n \frac{f(x_{i,t}) - f(x_t^B)}{\sum_{j=1}^{N} f(x_{i,t}) - f(x_t^B)} \right) \qquad (9.4)$$

Then the force, $F_{i,j}^t$, between two points $x_{i,t}$ and $x_{j,t}$ is calculated using:

$$F_{i,j}^t = \begin{cases} \left(x_{j,t} - x_{i,t} \right) \frac{q_{i,t} \cdot q_{j,t}}{\left\| x_{j,t} - x_{i,t} \right\|^2} & \text{if } f(x_{i,t}) > f(x_{j,t}) \\ \left(x_{i,t} - x_{j,t} \right) \frac{q_{i,t} \cdot q_{j,t}}{\left\| x_{j,t} - x_{i,t} \right\|^2} & \text{if } f(x_{i,t}) \leq f(x_{j,t}) \end{cases} \qquad (9.5)$$

The total force, F_i^t, corresponding to $x_{i,t}$ is now calculated as:

$$F_i^t = \sum_{j=1, j \neq i}^{N} F_{i,j}^t \tag{9.6}$$

Move the point $x_{i,t}$ along F_i^t (Line 5): In this step, each point $x_{i,t}$ except for x_t^B is moved along the total force F_i^t using:

$$x_{i,t} = x_{i,t} + \lambda \frac{F_i^t}{\|F_i^t\|}(RNG), \quad i = 1, 2, \ldots, N; \quad i \neq B \tag{9.7}$$

where $\lambda \sim Unif(0, 1)$ for each coordinate of $x_{i,t}$, and RNG denotes the allowed range of movement toward the lower or upper bound for the corresponding dimension.

Local search (Line 6): For each $y_{i,t}$ a maximum of $iter_{local}$ points are generated in each coordinate direction in the δ neighbourhood of $y_{i,t}$. This means that the process of generating local point is continued for each $y_{i,t}$ until either a better $z_{i,t}$ is found or the $n \times Iter_{local}$ trial is reached.

Selection for the next iteration (Line 7): In this step, $x_{i,t+1} \in \mathbf{Sp}_{t+1}$ are selected from $y_{i,t}$ and $z_{i,t}$ using Eq. (9.1), and the best point is identified using Eq. (9.2) for minimization or Eq. (9.3) for maximization.

As it can be seen from Eqs. (9.1)–(9.7), the process to compute the elements of the new population $\mathbf{Sp}_{t+1}$ involves several operations that consider local and global information. Such process is more laborious than most of the evolutionary approaches which use only one equation to modify the individual position. This fact could be considered as an implementation disadvantage of the EMO method.

9.3 Tsallis Entropy

The entropy is defined in thermodynamic to measure the order of irreversibility in the universe. The concept of entropy physically expresses the amount of disorder of a system [4, 8]. In information theory, Shannon redefines the theory proposed by Boltzmann-Gibbs, and employ the entropy to measure the uncertainty regarding information of a system [8]. In other words, is possible the quantitatively measurement of the amount of information produced by a process.

The entropy in a discrete system takes a probability distribution defined as $p = \{p_i\}$, which represents the probability of find the system in a possible state i. Notice that $0 \leq p_i \leq 1$, $\sum_{i=1}^{k} p_i = 1$, and k is the total number of states. In addition, a physical or information system can be decomposed in two statistical independent subsystems A and B with probabilities p^A and p^B, where the probability of the composed system is given by $p^{A+B} = p^A \cdot p^B$. Such definition has been verified using the extensive property (additive) (Eq. 9.8) proposed by Shannon [4, 8]:

$$S(A + B) = S(A) + S(B) \tag{9.8}$$

Tsallis proposes a generalized form of statistics based on the related concepts and the multi-fractal theory. The Tsallis entropic form is an important tool used to describe the thermo statistical properties of non-extensive systems and is defined as:

$$S_q = \frac{1 - \sum_{i=1}^{k} (p_i)^q}{q - 1} \tag{9.9}$$

where S is the Tsallis entropy, q is the Tsallis entropic index that represents the degree of non-extensivity and k is the total number of possibilities of the system. Since Tsallis entropy is non-extensive, it is necessary to redefine the additive entropic rule of Eq. (9.8).

$$S_q(A + B) = S_q(A) + S_q(B) + (1 - q) \cdot S_q(A) \cdot S_q(B) \tag{9.10}$$

Since image segmentation has non-additive information content, it is possible to use the Tsallis entropy to find the best thresholds [8]. A digital gray scale image has k gray levels that are defined by the histogram. The easiest thresholding considers to classes divided by a one threshold (bi-level), to solve this problem is considered the probability distribution of the gray levels ($p_i = p_1, p_2, \ldots p_k$). For each class A and B two probability distributions are created (see Eq. 9.11).

$$p_A = \frac{p_1}{P^A}, \frac{p_2}{P^A}, \ldots \frac{p_{th}}{P^A} \quad \text{and} \quad p_B = \frac{p_1}{P^B}, \frac{p_2}{P^B}, \ldots \frac{p_k}{P^B} \tag{9.11}$$

where:

$$P^A = \sum_{i=1}^{th} p_i \quad \text{and} \quad P^B = \sum_{i=th+1}^{k} p_i \tag{9.12}$$

The TE for class A and Class B is defined as follows:

$$S_q^A(th) = \frac{1 - \sum_{i=1}^{th} \left(\frac{p_i}{P^A}\right)^q}{q - 1}, \quad S_q^B(th) = \frac{1 - \sum_{i=th+1}^{k} \left(\frac{p_i}{P^B}\right)^q}{q - 1} \tag{9.13}$$

TE value depends directly on the selected threshold value, and it maximizes the information measured between two classes. If the value of $S_q(th)$ is maximized it means that th is the optimal value. In order to verify the efficiency of the selected th, in Eq. (9.14) is proposed an objective function using Eq. (9.10):

$$TH_{opt}(th) = \arg\max\left(S_q^A(th) + S_q^B(th) + (1 - q) \cdot S_q^A(th) \cdot S_q^B(th)\right) \tag{9.14}$$

The previous description of this bi-level method can be extended for the identification of multiple thresholds. Considering nt thresholds, it is possible separate the original image into (nt-1) classes. Under the segmentation approach, the optimization problem turns into a multidimensional situation. The candidate solutions are conformed as $\mathbf{th}^j = [th_1, th_2, \ldots th_{nt}]$. For each class is computed the entropy using the Tsallis methodology and the objective function is redefined as follows:

$$TH_{opt}(\mathbf{th}) = \arg\max\Big(S_q^1(th_1) + S_q^2(th_2) + \ldots$$

$$\cdots + S_q^{nt}(th_{nt}) + (1 - q) \cdot S_q^1(th_1) \cdot S_q^2(th_2) \cdot \ldots \cdot S_q^{nt}(th_{nt}) \Big) \quad (9.15)$$

where:

$$S_q^1(th_1) = \frac{1 - \sum_{i=1}^{th_1} \left(\frac{p_i}{P^1}\right)^q}{q - 1}, \quad S_q^2(th) = \frac{1 - \sum_{i=th_1+1}^{th_2} \left(\frac{p_i}{P^2}\right)^q}{q - 1}, \ldots$$

$$\ldots, S_q^{nt}(th) = \frac{1 - \sum_{i=th_{nt-1}+1}^{th_{nt}} \left(\frac{p_i}{P^{nt}}\right)^q}{q - 1} \quad (9.16)$$

Notice that for each threshold the entropy is computed and corresponds to a specific class. However there exist an extra class it means that exist $nt + 1$ classes. The extra class is considered *default* class because it is computed from nt to k (Eq. 9.17).

$$S_q^{def}(th_k) = \frac{1 - \sum_{i=th_{nt}+1}^{k} \left(\frac{p_i}{P^k}\right)^q}{q - 1} \quad (9.17)$$

From Eq. (9.14), it is evident that TE presents a simple functional formulation for bi-level thresholding. However, as it is shown by Eqs. (9.15) and (9.16), when it is considered multilevel thresholding (MT), its evaluation becomes computationally expensive, since each threshold point adds restrictions, multimodality and complexity to its functional formulation. Therefore, in the process of finding the appropriate threshold values, it is desired to limit the number of evaluations of the TE objective function. Under such circumstances, most of the optimization algorithms do not seem to be suited to face such problems as they usually require many evaluations before delivering an acceptable result.

9.4 Multilevel Thresholding Using EMO and Tsallis Entropy

This chapter introduces an algorithm for multilevel segmentation based on the Electromagnetism-Like algorithm (EMO). In this approach, the EMO algorithm is used to find the optimal threshold values by maximizing the complex Tsallis entropy. Different to other evolutionary methods, EMO exhibits interesting search capabilities such as fast convergence still keeping its ability to avoid local minima in high modality environments [10–12]. Meanwhile, some studies like [13–15] demonstrate that the EMO algorithm presents the best balance between optimization results and demand of function evaluations. As a result, the proposed segmentation algorithm can substantially reduce the number of function evaluations preserving the good search capabilities of an evolutionary method. However, as it can be seen from Eqs. (9.1)–(9.7), the process of EMO, to compute the elements of the new population, involves several operations that consider local and global information. Such process is more laborious than most of the evolutionary approaches which use only one equation to modify the individual position. This fact could be considered as an implementation disadvantage of the EMO method. In this section, the proposed approach is discussed.

9.4.1 Particle Representation

Each particle uses nt decision variables in the optimization algorithm. Such elements represent a different threshold point used for the segmentation. Therefore, the complete population is represented as:

$$\mathbf{Sp}_t = [\mathbf{th}_1, \mathbf{th}_2, \dots, \mathbf{th}_N], \quad \mathbf{th}_i = [th_1, th_2, \dots, th_{nt}]^T \tag{9.18}$$

where t represents the iteration number, T refers to the transpose operator, N is the size of the population.

9.4.2 EMO Implementation

The segmentation algorithm introduced in this chapter considers the Tsallis pseudo-additive entropic rule as objective function. The implementation of the EMO algorithm can be summarized into the following steps:

Step 1 Read the image I and store it into I_{Gr}.

Step 2 Obtain histogram h^{Gr} of I_{Gr}.

Step 3 Initialize the EMO parameters: $Iter_{\max}$, $Iter_{local}$, δ, k and N.

Step 4 Initialize a population $\mathbf{Sp}_t$ of N random particles with nt dimensions.

Step 5 Compute the Tsallis entropy $S_q^i(\mathbf{Sp}_t)$ for each element of $\mathbf{Sp}_t$, Eqs. (9.16) and (9.17). Evaluate $\mathbf{Sp}_t$ in the objective function $TH_{opt}(\mathbf{Sp}_t)$ Eq. (9.15).

Step 6 Compute the charge of each particle using Eq. (9.4), and with Eqs. (9.5) and (9.6) compute the total force vector.

Step 7 Move the entire population $\mathbf{Sp}_t$ along the total force vector using Eq. (9.7).

Step 8 Apply the local search to the moved population and select the best elements of this search based on their objective function values.

Step 9   The t index is increased in 1, If $t \geq Iter_{max}$ or if the stop criteria is satisfied the algorithm finishes the iteration process and jump to step 11. Otherwise jump to step 7.

Step 10 Select the particle that has the best x_t^{Bc} objective function value (Eqs. 9.3 and 9.15).

Step 11 Apply the thresholds values contained in x_t^{Bc} to the image I_{Gr}.

9.4.3   Multilevel Thresholding

Once the EMO algorithm finds the best threshold values that maximize the objective function. These are used to segment the image pixels. There exist several ways to apply the thresholds, in this chapter we use the following rule for two levels:

$$I_s(r, c) = \begin{cases} I_{Gr}(r, c) \text{ if } I_{Gr}(r, c) \leq th_1 \\ th_1 \quad\quad \text{ if } th_1 < I_{Gr}(r, c) \leq th_2 \\ I_{Gr}(r, c) \text{ if } I_{Gr}(r, c) > th_2 \end{cases} \qquad (9.19)$$

where $I_s(r, c)$ is the gray value of the segmented image, $I_{Gr}(r, c)$ is the gray value of the original image both in the pixel position r, c. th_1 and th_2 are the threshold values obtained by the EMO approach. Equation (9.19) can be easily extended for more than two levels (Eq. 9.20).

$$I_s(r, c) = \begin{cases} I_{Gr}(r, c) \text{ if } I_{Gr}(r, c) \leq th_1 \\ th_{i-1} \quad\quad \text{ if } th_{i-1} < I_{Gr}(r, c) \leq th_i , \quad i = 2, 3, \ldots nt - 1 \\ I_{Gr}(r, c) \text{ if } I_{Gr}(r, c) > th_{nt} \end{cases} \qquad (9.20)$$

9.5   Experimental Results

Like other approaches the implementation presented in this chapter has been tested using a set of 11 benchmark images. All the images have the same size (pixels) and they are in JPGE format. The TSEMO algorithms is also compared with methods as Cuckoo Search algorithm (CSA) [4] and Particle Swarm Optimization (PSO) [24]. Since all the methods are stochastic, it is necessary to employ statistical metrics to compare the efficiency of the algorithms. Hence, all algorithms are executed 35 times per image, according to the related literature the number the thresholds for test are $th = 2, 3, 4, 5$ [4, 24]. In each experiment the stop criteria is set to 50 iterations. In order to verify the stability at the end of each test the standard deviation (STD) is obtained (Eq. 9.21). If the STD value increases the algorithms becomes more instable [22].

$$STD = \sqrt{\sum_{i=1}^{Iter\max} \frac{(\sigma_i - \mu)}{Ru}} \tag{9.21}$$

In the same context the peak-to-signal ratio (PSNR) is also employed to compare the similarity of an image (image segmented) against a reference image (original image) based on the mean square error (MSE) of each pixel [4, 25, 26]. Both PSNR and MSE are defined as:

$$PSNR = 20 \log_{10}\left(\frac{255}{RMSE}\right), \quad (dB)$$

$$RMSE = \sqrt{\frac{\sum_{i=1}^{ro} \sum_{j=1}^{co} (I_{Gr}(i,j) - I_{th}(i,j))}{ro \times co}} \tag{9.22}$$

where I_{Gr} is the original image, I_{th} is the segmented image and ro, co are the total number of rows and columns of the image, respectively. The Structure Similarity Index ($SSIM$) is used to compare the structures of the original segmented image [27] it is defined in Eq. (9.23). A higher $SSIM$ value means that the performance of the segmentation is better.

$$SSIM(I_{Gr}, I_{th}) = \frac{(2\mu_{I_{Gr}}\mu_{I_{th}} + C1)(2\sigma_{I_{Gr}I_{th}} + C2)}{(\mu_{I_{Gr}}^2 + \mu_{I_{th}}^2 + C1)(\sigma_{I_{Gr}}^2 + \sigma_{I_{th}}^2 + C2)}$$

$$\sigma_{I_oI_{Gr}} = \frac{1}{N-1} \sum_{i=1}^{N} (I_{Gr_i} + \mu_{I_{Gr}})(I_{th_i} + \mu_{I_{th}}) \tag{9.23}$$

From Eq. (9.23) $\mu_{I_{Gr}}$ and $\mu_{I_{th}}$ are the mean value of the original and the thresholded image respectively, for each image the values of $\sigma_{I_{Gr}}$ and $\sigma_{I_{th}}$ corresponds to the standard deviation. $C1$ and $C2$ are constants used to avoid the instability when $\mu_{I_{Gr}}^2 + \mu_{I_{th}}^2 \approx 0$, experimentally in [4] both values are $C1 = C2 = 0.065$. Another method used to measure the quality of the segmented image is the Feature Similarity Index ($FSIM$) [28]. $FSIM$ calculates the similarity between two images, in this case the original gray scale image and the segmented image (Eq. 9.24). As $PSNR$ and $SSIM$ the higher value is interpreted as better performance of the thresholding method.

$$FSIM = \frac{\sum_{w\in\Omega} S_L(w)PC_m(w)}{\sum_{w\in\Omega} PC_m(w)} \tag{9.24}$$

From Eq. (9.24) Ω represents the entire domain of the image:

$$S_L(w) = S_{PC}(w)S_G(W)$$

$$S_{PC}(w) = \frac{2PC_1(w)PC_2(w) + T_1}{PC_1^2(w) + PC_2^2(w) + T_1}$$

Table 9.1 EMO parameters

$Iter_{max}$	$Iter_{local}$	δ	N
200	10	0.25	50

$$S_G(W) = \frac{2G_1(w)G_2(w) + T_2}{G_1^2(w) + G_2^2(w) + T_2} \tag{9.25}$$

G is the gradient magnitude (GM) of an image and is defined as $G = \sqrt{G_x^2 + G_y^2}$ and PC is the phase congruence:

$$PC(w) = \frac{E(w)}{\left(\varepsilon + \sum_n A_n(w)\right)} \tag{9.26}$$

From Eq. (9.26) $A_n(w)$ is the local amplitude on scale n and $E(w)$ is the magnitude of the response vector in w on n. ε is a small positive number and $PC_m(w) = \max(PC_1(w), PC_2(w))$. On the other hand, Table 9.1 presents the parameters for the EMO algorithm. They have been obtained using the criterion proposed in [9] and kept for all test images.

9.5.1 Result Using the Tsallis Entropy

This section presents the results obtained using the EMO algorithm to find the best thresholds using the Tsallis entropy for image segmentation. The approach is applied over the complete set of benchmark images whereas the results are registered in Table 9.2. Such results present the best threshold values obtained after testing the TSEMO algorithm, considering four different threshold points $th = 2, 3, 4, 5$. In Table 9.2, it is also shown the *PSNR*, *STD*, *SSIM* and *FSIM* values.

They have been selected five images of the set to show (graphically) the segmentation results. The selected images are presented in Fig. 9.1.

Table 9.3 shows the images obtained after processing 5 original images selected from the entire benchmark set, applying the proposed algorithm. The results present the segmented images considering four different threshold levels $th = 2, 3, 4, 5$. In Table 9.3, it is also shown the evolution of the objective function during one execution. From the results, it is possible to appreciate that the TSEMO converges around the first 100 iterations. The segmented images provide evidence that the outcome is better with $th = 4$ and $th = 5$; however, if the segmentation task does not requires to be extremely accurate then it is possible to select $th = 3$.

Comparisons of Tsallis Entropy

In order to demonstrate that the TSEMO is an interesting alternative for multilevel thresholding, the proposed algorithm is compared with two state-of-the-art imple-

Table 9.2 Result after applying the EMO and Tsallis entropy over the set of benchmark images

Image	k	Thresholds x_t^B	PSNR	STD	SSIM	FSIM
Camera man	2	71, 130	23.1227	31.00 E−04	0.9174	0.8901
	3	71, 130, 193	18.0122	72.01 E−04	0.8875	0.8456
	4	44, 84, 120, 156	24.9589	86.01 E−03	0.9363	0.9149
	5	44, 84, 120, 156, 196	23.0283	7.90 E−01	0.9289	0.8960
Lena	2	79, 127	23.9756	7.21 E−05	0.9083	0.8961
	3	79, 127, 177	21.0043	14.37 E−04	0.8660	0.8197
	4	62, 94, 127, 161	24.0020	18.69 E−03	0.9057	0.8851
	5	62, 94, 127, 161, 194	23.3736	39.82 E−02	0.8956	0.8684
Baboon	2	15, 105	23.5906	18.51 E−06	0.9480	0.9437
	3	51, 105, 158	19.9394	28.78 E−02	0.9011	0.9059
	4	33, 70, 107, 143	23.5022	22.65 E−02	0.9530	0.9594
	5	33, 70, 107, 143, 179	21.9540	37.13 E−01	0.9401	0.9417
Hunter	2	60, 119	22.8774	17.89 E−04	0.9192	0.8916
	3	60, 119, 179	20.2426	54.12 E−04	0.9031	0.8652
	4	46, 90, 134, 178	22.4723	1.94 E−02	0.9347	0.9159
	5	46, 90, 134, 178, 219	22.4025	1.23 E−01	0.9349	0.9173
Airplane	2	69, 125	25.4874	17.31 E−04	0.9685	0.9239
	3	69, 125, 180	22.9974	17.89 E−04	0.9433	0.8909
	4	55, 88, 122, 155	28.5400	19.21 E−03	0.9848	0.9677
	5	55, 88, 122, 155, 188	26.4997	35.08 E−03	0.9663	0.9417
Peppers	2	70, 145	19.6654	54.83 E−02	0.8697	0.8378
	3	70, 145, 223	17.2736	1.31 E−01	0.8437	0.7534
	4	46, 88, 132, 175	21.8275	3.02 E−04	0.8976	0.8552
	5	46, 88, 132, 175, 223	21.1207	6.34 E−03	0.8976	0.8304
Living room	2	55, 111	22.6665	47.11 E−03	0.9116	0.8966
	3	55, 111, 179	18.0379	15.27 E−04	0.8482	0.8132
	4	42, 85, 124, 162	21.7235	93.35 E−03	0.9170	0.9090
	5	42, 85, 124, 162, 201	21.3118	94.32 E−03	0.9183	0.9029
Blonde	2	62, 110	25.8389	31.91 E−04	0.9645	0.9503
	3	62, 110, 155	21.5001	37.05 E−04	0.9012	0.8759

(continued)

Table 9.2 (continued)

Image	k	Thresholds x_t^B	PSNR	STD	SSIM	FSIM
	4	36, 65, 100, 134	25.9787	17.45 E−03	0.9606	0.9491
	5	36, 65, 100, 134, 168	23.1835	48.20 E−03	0.9328	0.9077
Bridge	2	65, 131	20.1408	22.71 E−04	0.8619	0.8749
	3	65, 131, 191	18.7016	40.49 E−04	0.8410	0.8479
	4	45, 88, 131, 171	21.4247	38.48 E−03	0.9168	0.9279
	5	45, 88, 131, 171, 211	21.0157	66.16 E−03	0.9153	0.9217
Butterfly	2	83, 120	26.7319	96.11 E−03	0.9493	0.9195
	3	83, 120, 156	24.4582	39.04 E−03	0.9386	0.8934
	4	70, 94, 119, 144	27.0221	14.59 E−02	0.9653	0.9417
	5	70, 94, 119, 144, 172	25.7809	98.61 E−02	0.9610	0.9283
Lake	2	71, 121	27.8565	10.69 E−04	0.9729	0.9638
	3	71, 121, 173	23.7695	12.87 E−04	0.9399	0.9288
	4	41, 80, 119, 159	24.7454	11.97 E−03	0.9587	0.9422
	5	41, 80, 119, 159, 197	22.4347	11.80 E−03	0.9439	0.9213

mentations. The methods used for comparison are: the Cuckoo Search Algorithm (CSA) [29] and the Particle Swarm Optimization (PSO) [30]. Similar to the previous test, all the algorithms run 35 times over each selected image. The images used for this test are Camera man, Lena, Baboon, Hunter and Butterfly. For each image is computed the *PSNR*, *STD*, *SSIM*, *FSIM* values and the mean of the objective function.

The comparison results between the three methods are divided in two tables, Table 9.4 shows the *STD* and mean values of the fitness function. Table 9.5 presents the values of the quality metrics obtained after applying the thresholds over the test images.

The fitness values of four methods are statistically compared using a non-parametric significance proof known as the Wilcoxon's rank test [31, 32] that is conducted with 35 independent samples. Such proof allows assessing result differences among two related methods. The analysis is performed considering a 5% significance level over the best fitness (Tsallis entropy) value data corresponding to the five threshold points. Table 9.6 reports the p-values produced by Wilcoxon's test for a pair-wise comparison of the fitness function between two groups formed as TSEMO versus CSA, TSEMO versus PSO. As a null hypothesis, it is assumed that there is no difference between the values of the two algorithms tested. The alternative hypothesis considers an existent difference between the values of both approaches. All p-values reported in Table 9.6 are lower than 0.05 (5% significance level) which

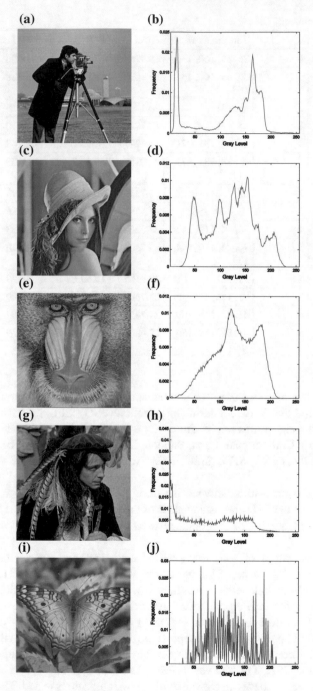

Fig. 9.1 **a** Camera man, **c** Lena, **e** Baboon, **g** Hunter and **i** Butterfly, the selected benchmark images. **b**, **d**, **f**, **h**, **j** histograms of the images

Table 9.3 Results after applying the EMO using Tsallis entropy over the selected benchmark images

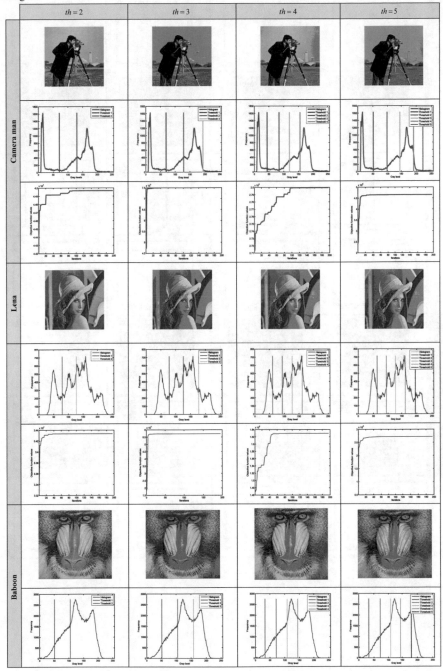

(continued)

Table 9.3 (continued)

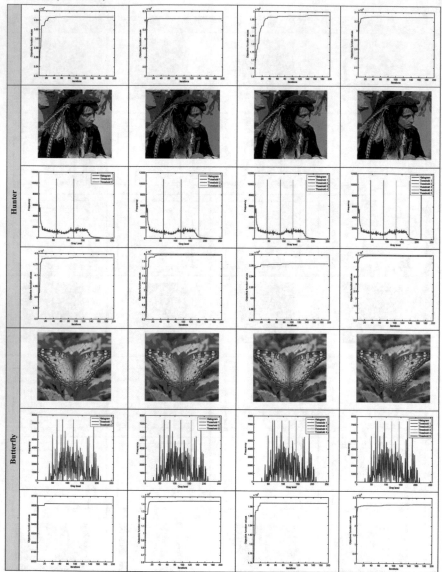

Table 9.4 Comparison of the *STD* and mean values of the TSEMO, CSA and PSO applied over the selected test images using Tsallis method

Image	k	TSEMO		CSA		PSO	
		STD	Mean	*STD*	Mean	*STD*	Mean
Camera man	2	31.00 E−04	4.49 E+04	89.56 E−04	4.02 E+04	83.00 E02	4.19 E+04
	3	72.01 E−04	7.49 E+04	98.32 E−04	6.99 E+04	89.00 E+00	7.27 E+04
	4	86.01 E−03	2.79 E+06	18.68 E−03	2.18 E+06	12.35 E+02	2.37 E+06
	5	7.90 E−01	4.65 E+06	69.98 E−01	4.56 E+06	5.38 E+03	4.28 E+06
Lena	2	7.21 E−05	3.43 E+04	2.61 E+00	3.33 E+04	15.27 E+00	3.30 E+04
	3	14.37 E−04	5.72 E+04	3.39 E+00	5.67 E+04	3.31 E+00	5.62 E+04
	4	18.69 E−03	1.62 E+06	5.52 E+00	1.45 E+06	7.35 E+00	1.45 E+06
	5	39.82 E−02	2.71 E+06	8.50 E+01	2.55 E+06	2.92 E+00	2.59 E+06
Baboon	2	18.51 E−06	3.64 E+04	15.11 E−02	3.47 E+04	2.64 E+00	3.40 E+04
	3	28.78 E−02	6.08 E+04	40.80 E−02	6.05 E+04	1.44 E+00	6.03 E+04
	4	22.65 E−02	1.97 E+06	62.02 E−02	1.90 E+06	8.11 E+00	1.86 E+06
	5	37.13 E−01	3.29 E+06	52.74 E−02	3.20 E+06	2.68 E+00	3.20 E+06
Hunter	2	17.89 E−04	4.78 E+04	7.38 E−04	4.70 E+04	4.38 E+00	4.72 E+04
	3	54.12 E−04	7.97 E+04	2.95E−04	7.89 E+04	9.47 E+00	7.85 E+04
	4	1.94 E−02	2.96 E+06	1.62 E−01	2.93 E+06	1.04 E+01	2.92 E+04
	5	1.23 E−01	4.94 E+06	2.46 E−01	4.89 E+06	3.23 E+02	4.75 E+04
Butterfly	2	96.11 E−03	8.61 E+03	12.78 E−02	8.56 E+03	6.36 E−01	8.55 E+03
	3	39.04 E−03	1.43 E+04	19.00 E−02	1.38 E+04	11.56 E−01	1.35 E+04
	4	14.59 E−02	1.88 E+05	11.04 E−01	1.80 E+05	1.04 E+00	1.81 E+05
	5	98.61 E−02	3.14 E+05	1.58 E+00	3.07 E+05	3.58 E+00	2.96 E+05

Table 9.5 Comparison of the *PSNR*, *SSIM* and *FSIM* values of the TSEMO, CSA and PSO applied over the selected test images using Tsallis method

Image	k	TSEMO			CSA			PSO		
		PSNR	*SSIM*	*FSIM*	*PSNR*	*SSIM*	*FSIM*	*PSNR*	*SSIM*	*FSIM*
Camera man	2	23.1227	0.9174	0.8901	23.1194	0.9173	0.8901	22.9737	0.9160	0.8871
	3	18.0998	0.8875	0.8509	18.7480	0.8918	0.8456	18.0122	0.8874	0.8441
	4	25.0021	0.9369	0.9151	24.5479	0.9349	0.9097	23.3230	0.9280	0.8976
	5	22.9136	0.9286	0.8950	22.5284	0.9243	0.8891	21.9598	0.9222	0.8839
Lena	2	23.9982	0.9088	0.8966	23.9756	0.9083	0.8961	23.9594	0.9085	0.8953
	3	21.2592	0.8699	0.8255	20.9669	0.8655	0.8192	20.9989	0.8659	0.8196
	4	23.9783	0.9056	0.8849	23.9493	0.9056	0.8846	23.8175	0.9032	0.8815
	5	23.4275	0.8954	0.8691	23.3099	0.8960	0.8689	23.3777	0.8949	0.8674
Baboon	2	23.7510	0.9496	0.9452	23.5906	0.9480	0.9410	23.5048	0.9475	0.9323
	3	19.9386	0.9007	0.9057	19.9031	0.8810	0.8759	19.8021	0.8729	0.8729
	4	23.5165	0.9532	0.9593	23.5106	0.9270	0.9295	23.5163	0.9125	0.9159
	5	22.0538	0.9410	0.9408	21.9071	0.9399	0.9112	21.7165	0.9350	0.9377
Hunter	2	22.8783	0.9192	0.8916	22.8074	0.9089	0.8826	22.7910	0.9093	0.8818
	3	20.2581	0.9034	0.8654	20.0026	0.8931	0.8552	20.0858	0.8921	0.8521
	4	22.4221	0.9341	0.9159	21.3972	0.9237	0.9055	21.5061	0.9244	0.9024
	5	22.5014	0.9355	0.9199	21.3171	0.9236	0.9063	21.3754	0.9254	0.9005
Butterfly	2	26.8352	0.9504	0.9212	25.7319	0.9493	0.9195	25.1635	0.9431	0.9150
	3	24.4144	0.9383	0.8926	23.4545	0.9300	0.8834	23.5251	0.9315	0.8846
	4	27.1226	0.9653	0.9420	26.0314	0.9653	0.9317	26.0810	0.9653	0.9321
	5	25.8838	0.9609	0.9285	24.0086	0.9516	0.9201	24.4870	0.9533	0.9142

is a strong evidence against the null hypothesis, indicating that the TSEMO fitness values for the performance are statistically better and it has not occurred by chance.

On the other hand, to compare the fitness of the three methods Table 9.7 shows the fitness values obtained for the reduced set of image (5 images). Each algorithm runs 1000 times and the best value of each run is stored, at the end of the evolution process the best stored values are plotted. From Table 9.7 it is possible to analyse that TSEMO and CSA reach the maximum entropy values in less iterations than the PSO method.

9.6 Summary

This chapter explains the implementation of the EMO algorithm for image segmentation using the Tsallis entropy. For this purpose, the segmentation process is considered as an optimization problem where EMO is employed to find the optimal

Table 9.6 Wilcoxon p-values of the compared algorithm TSEMO versus CSA and TSEMO versus PSO

Image	k	p-values	
		TSEMO versus CS	TSEMO versus PSO
Camera man	2	6.2137 E−07	8.3280 E−06
	3	1.0162 E−07	2.0000 E−03
	4	8.8834 E−08	13.710 E−03
	5	16.600 E−03	50.600 E−03
Lena	2	3.7419 E−08	1.6604 E−04
	3	1.4606 E−06	1.3600 E−02
	4	1.2832 E−07	2.9000 E−03
	5	3.9866 E−05	8.9000 E−03
Baboon	2	1.5047 E−06	2.5500 E−02
	3	6.2792 E−05	5.1000 E−03
	4	2.1444 E−12	3.3134 E−05
	5	2.1693 E−11	1.8000 E−03
Hunter	2	2.2100 E−02	2.2740 E−02
	3	3.6961 E−04	1.1500 E−02
	4	6.8180 E−02	9.9410 E−09
	5	5.8200 E−02	2.4939 E−04
Airplane	2	3.0000 E−03	6.6300 E−03
	3	7.6000 E−03	3.5940 E−02
	4	4.8092 E−12	1.1446 E−06
	5	1.0023 E−09	2.7440 E−02
Peppers	2	2.7419 E−04	1.3194 E−04
	3	2.6975 E−08	3.5380 E−02
	4	1.5260 E−08	6.0360 E−02
	5	7.2818 E−08	7.6730 E−02
Living room	2	1.4000 E−03	2.6340 E−02
	3	6.8066 E−08	2.8000 E−03
	4	8.7456 E−07	5.8730 E−03
	5	1.7000 E−03	5.1580 E−03
Blonde	2	3.0000 E−03	4.1320 E−02
	3	5.9000 E−03	8.9300 E−02
	4	1.3800 E−02	2.7700 E−02
	5	2.3440 E−02	5.6000 E−03
Bridge	2	1.5000 E−03	1.5700 E−02
	3	1.4300 E−02	1.5350 E−02
	4	1.7871 E−06	7.0400 E−03

(continued)

Table 9.6 (continued)

Image	k	p-values	
		TSEMO versus CS	TSEMO versus PSO
	5	8.7000 E−03	1.2400 E−02
Butterfly	2	1.5000 E−03	1.1150 E−02
	3	3.1800 E−02	1.3760 E−02
	4	4.8445 E−07	8.1800 E−03
	5	1.6000 E−02	1.0630 E−02
Lake	2	7.6118 E−06	2.9500 E−02
	3	1.2514 E−06	6.5644 E−06
	4	2.2366 E−10	6.6000 E−03
	5	5.3980 E−06	9.4790 E−03

Table 9.7 Fitness comparison of PSO (blue line), CSA (black line) and EMO (red line) applied for multilevel thresholding using TE

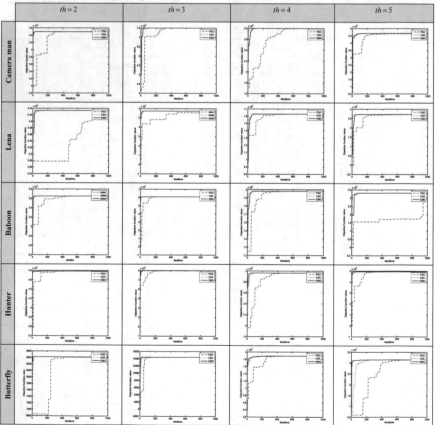

threshold points that maximize the Tsallis entropy (TE). In this approach, the algorithm uses particles that represents the candidate solutions conformed by thresholds. The TE evaluates the segmentation quality of such particles. Guided by the values of the Tsallis entropy, the set of candidate solutions are modified by the EMO operators so that they can improve their segmentation quality as the optimization process evolves.

In order to evaluate the quality of the segmented images, the use of the *PSNR*, *STD*, *SSIM* and *FSIM* is proposed. Such metrics considers the coincidences between the original and the segmented image. The study compares the proposed approach with other two similar approaches the Cuckoo Search algorithm (CSA) and Particle Swarm Optimization algorithm (PSO). The efficiency of the algorithms is evaluated in terms of *PSNR*, *STD*, *SSIM*, *FSIM* and fitness values. Such comparisons provide evidence of the accuracy, convergence and robustness of the proposed approach. The fitness of TSEMO is compared with the CSA and PSO where is possible to see that both EMO and CSA need a reduced number of iterations to converge. However, the speed of convergence of EMO is higher than de CSA in the same way PSO is the slower and it has lack of accuracy. Although the results offer evidence to demonstrate that the TSEMO method can yield good results on complicated images, the aim of our chapter is not to devise a multilevel thresholding algorithm that could beat all currently available methods, but to show that electro-magnetism systems can be effectively considered as an attractive alternative for this purpose.

References

1. Otsu N (1979) A threshold selection method from gray-level histograms. IEEE Trans Syst Man Cybern 9:62–66. https://doi.org/10.1109/TSMC.1979.4310076
2. Kapur JN, Sahoo PK, Wong AKC (1985) A new method for gray-level picture thresholding using the entropy of the histogram. Comput Vis Graph Image Process 29:273–285
3. Sathya PD, Kayalvizhi R (2011) Optimal multilevel thresholding using bacterial foraging algorithm. Expert Syst Appl 38:15549–15564. https://doi.org/10.1016/j.eswa.2011.06.004
4. Agrawal S, Panda R, Bhuyan S, Panigrahi BK (2013) Tsallis entropy based optimal multilevel thresholding using cuckoo search algorithm. Swarm Evol Comput 11:16–30. https://doi.org/10.1016/j.swevo.2013.02.001
5. Tsallis C (1988) Possible generalization of Boltzmann-Gibbs statistics. J Stat Phys 52:479–487. https://doi.org/10.1007/BF01016429
6. Tang EK, Suganthan PN, Yao X (2006) An analysis of diversity measures. Mach Learn 65:247–271. https://doi.org/10.1007/s10994-006-9449-2
7. Tsallis C (2002) Entropic nonextensivity: a possible measure of complexity. Chaos Solitons Fractals 13:371–391. https://doi.org/10.1016/S0960-0779(01)00019-4
8. Zhang Y, Wu L (2011) Optimal multi-level thresholding based on maximum Tsallis entropy via an artificial bee colony approach. Entropy 13:841–859. https://doi.org/10.3390/e13040841
9. Birbil ŞI, Fang SC (2003) An electromagnetism-like mechanism for global optimization. J Glob Optim 25:263–282. https://doi.org/10.1023/A:1022452626305
10. De Castro LN, Von Zuben FJ (2002) Learning and optimization using the clonal selection principle. IEEE Trans Evol Comput 6:239–251. https://doi.org/10.1109/TEVC.2002.1011539
11. De Jong K (1988) Learning with genetic algorithms: an overview. Mach Learn 3:121–138. https://doi.org/10.1007/BF00113894

12. Dorigo M, Maniezzo V, Colorni A (1996) The ant systems: optimization by a colony of cooperative agents. IEEE Trans Man Mach Cybern B 26
13. Birbil ŞI, Fang SC, Sheu RL (2004) On the convergence of a population-based global optimization algorithm. J Glob Optim 30:301–318
14. Rocha AMAC, Fernandes EMGP (2009) Modified movement force vector in an electromagnetism-like mechanism for global optimization. Optim Methods Softw 24:253–270
15. Fernandes EMGP, Rocha AMAC (2009) Hybridizing the electromagnetism-like algorithm with descent search for solving engineering design problems. Int J Comput Math 86:1932–1946. https://doi.org/10.1080/0020716YYxxxxxxxx
16. Wu P, Yang W-H, Wei N-C (2004) An electromagnetism algorithm of neural network analysis—an application to textile retail operation. J Chinese Inst Ind Eng 21:59–67. https://doi.org/10.1080/10170660409509387
17. Naderi B, Tavakkoli-Moghaddam R, Khalili M (2010) Electromagnetism-like mechanism and simulated annealing algorithms for flowshop scheduling problems minimizing the total weighted tardiness and makespan. Knowl-Based Syst 23:77–85. https://doi.org/10.1016/j.knosys.2009.06.002
18. Hung HL, Huang YF (2011) Peak to average power ratio reduction of multicarrier transmission systems using electromagnetism-like method. Int J Innov Comput Inf Control 7:2037–2050
19. Yurtkuran A, Emel E (2010) A new hybrid electromagnetism-like algorithm for capacitated vehicle routing problems. Expert Syst Appl 37:3427–3433. https://doi.org/10.1016/j.eswa.2009.10.005
20. Jhang JY, Lee KC (2009) Array pattern optimization using electromagnetism-like algorithm. AEU—Int J Electron Commun 63:491–496. https://doi.org/10.1016/j.aeue.2008.04.001
21. Lee CH, Chang FK (2010) Fractional-order PID controller optimization via improved electromagnetism-like algorithm. Expert Syst Appl 37:8871–8878. https://doi.org/10.1016/j.eswa.2010.06.009
22. Ghamisi P, Couceiro MS, Benediktsson JA, Ferreira NMF (2012) An efficient method for segmentation of images based on fractional calculus and natural selection. Expert Syst Appl 39:12407–12417. https://doi.org/10.1016/j.eswa.2012.04.078
23. Rocha AMAC, Fernandes EMGP (2009) Hybridizing the electromagnetism-like algorithm with descent search for solving engineering design problems. Int J Comput Math 86:1932–1946. https://doi.org/10.1080/00207160902971533
24. Sri N, Raja M, Kavitha G, Ramakrishnan S (2012) Analysis of vasculature in human retinal images using particle swarm optimization based Tsallis multi-level thresholding and similarity measures. Lect Notes Comput Sci (including Subser Lect Notes Artif Intell Lect Notes Bioinformatics) 7677:380–387
25. Horng M-H (2011) Multilevel thresholding selection based on the artificial bee colony algorithm for image segmentation. Expert Syst Appl 38:13785–13791. https://doi.org/10.1016/j.eswa.2011.04.180
26. Oh I-S, Lee J-S, Moon B-R (2004) Hybrid genetic algorithms for feature selection. IEEE Trans Pattern Anal Mach Intell 26:1424–1437. https://doi.org/10.1109/TPAMI.2004.105
27. Wang Z, Bovik AC, Sheikh HR, Simoncelli EP (2004) Image quality assessment: from error visibility to structural similarity. IEEE Trans Image Process 13:600–612. https://doi.org/10.1109/TIP.2003.819861
28. Zhang Lin, Zhang Lei, XuanqinMou DZ (2011) FSIM: a feature similarity index for image. IEEE Trans Image Process 20:2378–2386
29. Yang X-S, Deb S (2010) Cuckoo search via levy flights. 210–214. https://doi.org/10.1109/NABIC.2009.5393690
30. Kennedy J, Eberhart RC (1995) Particle swarm optimization. In: 1995 Proceedings of IEEE international conference on neural networks, vol 4, pp 1942–1948. https://doi.org/10.1109/ICNN.1995.488968

31. Wilcoxon F (1945) Individual comparisons by ranking methods. Biom Bull 1:80. https://doi.org/10.2307/3001968
32. García S, Molina D, Lozano M, Herrera F (2009) A study on the use of non-parametric tests for analyzing the evolutionary algorithms. J Heuristics 15:617–644

Chapter 10
Image Segmentation with Minimum Cross Entropy

10.1 Introduction

Image segmentation is a crucial process involved in tasks such as object tracking, detection of regions of interest, medical imaging, among others [1–3]. Its purpose is the partition of the image into homogeneous classes. The elements of each class will share common properties such as intensity or texture. Image Thresholding (TH) uses the gray-scale histogram to select threshold intensity values which are meant to separate classes. Bi-level thresholding is the simplest case and only uses one threshold value to create two classes. This approach extracts an object from its background. While bi-level thresholding is easy to implement, Multilevel Thresholding (MTH) can provide more information while generating a finite number of classes. Segmentation methods based on threshold can be divided into parametric and nonparametric [4–6]. Parametric approaches estimate parameters of a probability density function to describe each class, but this approach is computationally expensive. By contrast, nonparametric approaches use criteria such as between-class variance, entropy and error rate [7–9]. These criteria are optimized to find the optimal threshold value providing robust and accurate methods [10].

The development of information theory provides opportunities to explore the use of several entropies to find efficient schemes intended to separate objects and its background. Kapur entropy [8], Tsallis entropy [11], and cross entropy [12] to list some. The Minimum Cross Entropy Thresholding (MCET) algorithm is widely used in the literature to segment images. Li and Lee proposed a method that identifies thresholds by minimizing the cross entropy between the original and segmented images [13]. Yin proposed a recursive programming technique which reduce the magnitude of computing the objective function of the MCET [14]. As an alternative to parametric techniques, the problem of MTH has also been handled through evolutionary computation techniques (ECT). Such approaches produced several MTH applications by selecting different evolutionary computation techniques and optimizing various criteria, such as Particle Swarm Optimization (PSO), Firefly (FF)

© Springer Nature Switzerland AG 2019
D. Oliva et al., *Metaheuristic Algorithms for Image Segmentation: Theory and Applications*, Studies in Computational Intelligence 825, https://doi.org/10.1007/978-3-030-12931-6_10

and Cross-Entropy [15], Electromagnetism-Like Optimization (EMO) and Tsallis entropy [16], whose results have been individually reported.

This chapter introduces the MCET-CSA for image thresholding. The MCET-CSA is based on the recently published Evolutionary Computation Technique (ECT) known as Crow Search Algorithm (CSA) in conjunction with the Minimum Cross Entropy Thresholding algorithm (MCET). Multilevel Thresholding partitions the image into a finite number of classes by the determination of threshold values, where every new threshold increases the complexity of the problem by including more restrictions and the also modifying the modality of the search space, especially when the histogram of the image presents irregularities. The experiments are performed considering an entropy-based criterion as objective functions; the cross entropy. Thus, CSA is used to minimize the cross entropy among classes. In order to assess the performance of MCET-CSA, two state-of-the-art ECT are implemented for the same application; the Differential Evolution (DE) and the Harmony Search (HS) [17, 18].

The proposed methodology has been validated using a set of images extracted from Brainweb database (http://brainweb.bic.mni.mcgill.ca/brainweb). The first benchmark set is used to analyze the search capabilities of CSA. For this purpose, a high-dimensional framework is used to demonstrate the search capabilities of the algorithm. Most related literature focus on the search for optimal threshold values in an up to 5-dimensional scheme (5 threshold values), while in this chapter, the search is conducted in an up to 32-dimensional space. The aim of this set of benchmark images is to determine the quality of segmented MRI brain images. For both benchmarks, MCET-CSA obtained results showing competitive segmentation quality and consistency. Moreover, results are validated within a statistically significant framework.

The remainder of the chapter is organized as follows: Sect. 10.2 describes cross entropy. In Sect. 10.3 the standard version of CSA is introduced. Section 10.4 describes the implementation of the proposed algorithm. Section 10.5 discusses the experimental results after testing the MCET-CSA, and a statistical analysis is presented. Lastly, the work is concluded in Sect. 10.6.

10.2   Minimum Cross Entropy

Thermodynamics defines entropy as a metric to measure the order of irreversibility in the universe. In physical terms, entropy expresses the amount of disorder of a system. In the context of information theory, various formulations of entropy are used in order to measure homogeneity of data. Kullback proposed the cross entropy in [19]. Let $\mathbf{J} = \{j_1, j_2, \ldots, j_N\}$ and $\mathbf{G} = \{g_1, g_2, \ldots, g_N\}$ be two probability distributions on the same set. The cross entropy between $\mathbf{F}$ and $\mathbf{G}$ is an information theoretic distance between the two distributions and it is defined by:

$$D(\mathbf{J}, \mathbf{G}) = \sum_{i=1}^{N} j_i \log \frac{j_i}{g_i} \tag{10.1}$$

The minimum cross entropy thresholding (MCET) algorithm [13] selects the threshold by minimizing the cross entropy between the thresholded version and its original image. The original image is $\mathbf{I}$ and $h^{Gr}(i)$, $i = 1, 2, \ldots, L$, is the corresponding histogram with L being the number of gray levels. Then the thresholded image, denoted by $\mathbf{I}_t$ using th as the threshold value is constructed by:

$$\mathbf{I}_t(x, y) = \begin{cases} \mu(1, th), & \mathbf{I}(x, y) < th, \\ \mu(th, L+1), & \mathbf{I}(x, y) \geq th, \end{cases} \tag{10.2}$$

where:

$$\mu(a, b) = \sum_{i=a}^{b-1} i h^{Gr}(i) / \sum_{i=a}^{b-1} h^{Gr}(i) \tag{10.3}$$

Since Eq. (10.2) generates a thresholded image instead of an entropy value, the cross entropy is rewritten as an objective function defined in Eq. (10.4).

$$f_{Cross}(th) = \sum_{i-1}^{th-1} i h^{Gr}(i) \log\left(\frac{i}{\mu(1, th)}\right) + \sum_{i=th}^{L} i h^{Gr}(i) \log\left(\frac{i}{\mu(th, L+1)}\right) \tag{10.4}$$

The objective function considers a single threshold value for a bilevel thresholding. Equation (10.4) can be extended to a multilevel approach. First Eq. (10.4) can be expressed as:

$$f_{Cross}(th) = \sum_{i=1}^{L} i h^{Gr}(i) \log(i) - \sum_{i=1}^{th-1} i h^{Gr}(i) \log(\mu(1, th)) - \cdots - \sum_{i=th}^{L} i h^{Gr}(i) \log(\mu(th, L+1))$$
$$\tag{10.5}$$

The multilevel approach is based on the use of the vector $\mathbf{th} = [th_1, th_2, \ldots, th_{nt}]$ which contains nt different thresholds values. Equation (10.6) presents the objective function for multiple thresholds.

$$f_{Cross}(\mathbf{th}) = \sum_{i=1}^{L} i h^{Gr}(i) \log(i) - \sum_{i=1}^{nt} H_i \tag{10.6}$$

Where k is the number of threshold and entropies to calculate and is defined as:

$$H_1 = \sum_{i=1}^{th_1-1} ih^{Gr}(i)\log(\mu(1, th_1))$$

$$H_k = \sum_{i=th_{k-1}}^{th_k-1} ih^{Gr}(i)\log(\mu(th_{k-1}, th_k)), \ 1 < k < nt \qquad (10.7)$$

$$H_{nt} = \sum_{i=th_{nt}}^{L} ih^{Gr}(i)\log(\mu(th_{nt}, L+1))$$

10.3   The Crow Search Algorithm

The Crow Search Algorithm proposed by Askarzadeh [20], is an evolutionary method inspired by the intelligent behavior of crows. Crows are considered to be among the world's most intelligent animals [21], thus, their behaviors can provide interesting heuristics. CSA is inspired by the thievery behavior that crows exhibit; such behavior can be summarized as follows: crows memorize the position where they hide food surplus, a crow can follow another one to do thievery on their caches, in order to protect their food hideouts crows can mislead thieves by moving randomly. From a computational point of view, in the CSA the population $\mathbf{C}^k(\{\mathbf{c}_1^k, \mathbf{c}_2^k, \ldots, \mathbf{c}_N^k\})$ of N individuals (crows) is evolved from initial point ($k = 0$) to a total number of iterations ($k = gen$). Each crow $\mathbf{c}_i^k(i \in [1, \ldots, N])$ represents a d-dimensional vector $\{\mathbf{c}_{i,1}^k, \mathbf{c}_{i,2}^k, \ldots, \mathbf{c}_{i,N}^k\}$ where each dimension corresponds to a decision variable of the problem to be solved. In CSA, a new population $\mathbf{C}^{k+1}$ is generated considering two states: the first is when the crow is aware that is being followed and the second is when the crow is not aware. An awareness probability factor AP_i^k determines which state is selected. Each new element can be computed as follows:

$$\mathbf{c}_i^{k+1} = \begin{cases} \mathbf{c}_i^k + r_i \times fl \times \left(\mathbf{m}_j^k - \mathbf{c}_i^k\right) & r_j \geq AP_i^k \\ \text{random position} & \text{otherwise} \end{cases} \qquad (10.8)$$

where r_i and r_j are random numbers drawn from a uniform distribution between 0 and 1, fl is a parameter that controls the flight length. $\mathbf{m}_j^k$ is the memory of the crow j at iteration k; it stores the best position found so far by the crow j. CSA only require the configuration of two parameters, fl and AP_i^k. Besides, the update expression is quite simple, providing a user-friendly yet powerful implementation.

The required steps of CSA are presented in the flowchart of Fig. 10.1. The first stage is the initialization of the problem (dimensions and limits) and the parameter of the CSA (*fl, AP*, stop criteria). The next stages consist in randomly initialize the crow positions and evaluate them in the objective function. The positions are uniformly distributed in the search space. After that the new positions are generated according with Eq. (10.8) and their feasibility is verified. All the new positions are evaluated in the objective function and the memory is updated. Finally, the stop criteria is verified in order to terminate or continue with the iterative process. The stop criteria depends on the implementation of the CSA but two common rules are used: (1) the use of a

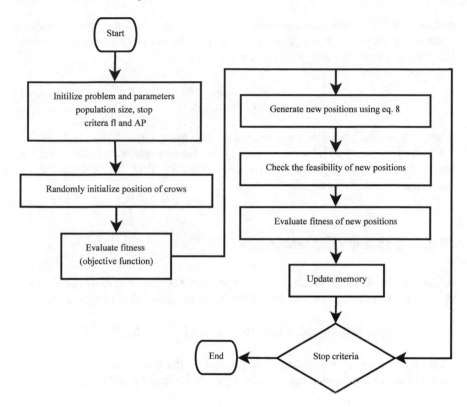

Fig. 10.1 Crow search algorithm

predefined number of iterations and (2) when the objective function value of the best crow converges and doesn't change along the iterations.

10.4 Minimum Cross Entropy and Crow Search Algorithm

This chapters aims to presents the implementation of the crow search algorithm for image thresholding using the minimum cross entropy. The multilevel cross entropy thresholding partitions the image into a finite number of classes by the determination of threshold values. Where every new threshold adds complexity to the problem by increasing the restrictions and the modality of the search space, especially when the histogram of the image presents irregularities. CSA is used to minimize the cross entropy among classes. At each generation, CSA encodes a set of candidate threshold points into a solution. The objective function uses the cross entropy to determinate the quality of the proposed solution. Following the rules of the CSA and the value of the objective function, new candidate solutions are generated using the predefined operators of CSA while improving the segmentation quality as the process evolves. Since multilevel thresholding can be treated as an optimization problem, the objective function for the cross-entropy criterion is stated as:

$$\arg \min_{\mathbf{th}} \ f_{Cross}(\mathbf{th})$$
$$\text{subject to} \quad \mathbf{th} \in \mathbf{X} \tag{10.9}$$

where $f_{Cross}(\mathbf{th})$ is the cross entropy function (Eq. 10.6) and $\mathbf{X} = \{th \in \mathbb{R}^{nt} | 0 \leq th_i \leq 255, i = 1, 2, \ldots, nt\}$ is the constrained feasible region, bounded by the intensity values of the interval 0–255. Thus, The CSA is used to find threshold values that solves the Eq. 10.6.

10.4.1 Solutions Representation

All the thresholds th are incorporated into the optimization algorithm as a decision variable in each element of the population. Thus, the population is represented as:

$$\mathbf{Sp}_t = [\mathbf{th}_1, \mathbf{th}_2, \ldots, \mathbf{th}_N], \mathbf{th}_i = [th_1, th_2, \ldots, th_{nt}]^T \tag{10.10}$$

where t is the iteration number, N is the size of the population, T refers to the transpose operator and nt is the number of thresholds applied to the image.

10.4.2 Crow Search Implementation

The implementation of the CSA presented in this chapter consider the minimum cross entropy as an objective function. The entire process of for image segmentation is summarized in the following steps:

Step 1   Read the image $\mathbf{I}$ and store it into $\mathbf{I}_{Gr}$.

Step 2   Calculate histogram h^{Gr} of $\mathbf{I}_{Gr}$.

Step 3   Initialize CSA parameters: Awareness probability AP_i^k and flight length fl.

Step 4   Initialize population $\mathbf{Sp}_t$ of N *random* particles with nt dimensions.

Step 5   Evaluate objective function (f_{Cross}) for each element of $\mathbf{Sp}_t$. Equation (10.6).

Step 6   Generate a random number r_j

Step 7   If $r_j \geq AP$ update solution via $\mathbf{c}_i^{k+1} = \mathbf{c}_i^k + r_i \times fl \times \left(\mathbf{m}_j^k - \mathbf{c}_i^k\right)$.

Step 8   If $r_j < AP$ update solution via $\mathbf{c}_i^{k+1} = $ random position

Step 9   Verify if all the N crows are feasible solutions.
Step 10   If the stop criteria are not satisfied jump to step 5.
Step 11   Generate segmented image $\mathbf{I}_s$ with $\mathbf{g}$ and Eq. (10.2).

The best solution obtained by the CSA contains the thresholds that according to the MCET provide the optimal segmentation of the image. The process for applying the thresholds to the histogram of the input image has been described in Chap. 9 and it could be used also for the values obtained by the CSA.

10.5 Experimental Results

The methodology presented in this chapter is validated using a set of brain magnetic resonance (MRI) images. A set of 8 images extracted from Brainweb database (http://brainweb.bic.mni.mcgill.ca/brainweb). This set is used to determine the performance of MCET-CSA in complex scenes. All experiments were done using Matlab 8.3 on an i5-4210 CPU @ 2.3 GHz with 6 GB of RAM.

Each experiment of a population based algorithm, such as CSA and DE, has a stop criterion set to 2500 iterations, but the algorithm is terminated if the fitness value is not improved during 10% of the stop criterion. Since HS is based on a single particle, its stop criterion is set to the size of the population of CSA and DE multiplied by 2500. Using this criterion the HS is able to find the best value [22].

To evaluate the quality of the segmented images provided by the CSA they are used the Standard Deviation (STD), the Peak-to-Signal Noise Ration (PSNR), the Structure Similarity Index and the Feature Similarity Index (FSIM). All these metrics were described in Sect. 9.5 in Chap. 9.

10.5.1 Parameters Setting

In this chapter for the CSA implementation the parameters used are the ones proposed on its own definition [20]. DE uses the parameters presented in [23] while HS uses [24] from a similar application. Table 10.1 presents all parameter for every algorithm employed in this chapter.

10.5.2 Segmentation of Brain Images from Magnetic Resonance

This subsection analyzes the results provided by MCET implementations based on CSA, DE, and HS, after being used to segment the medical benchmark images. The MCET-CSA is applied to the complete benchmark image set. The benchmark images

Table 10.1 Selected parameters for CSA, DE and HS

CSA	DE	HS
– Population N: 20 – Awareness probability AP:0.1 – Flight length fl:2	– Population N: 20 – Crossover probability: 0.7455 – Differential weight: 0.9362	– Harmony memory: 100 – Consideration rate: 0.75 – Pitch adjusting rate: 0.5 – Distance bandwidth: 0.5 – Number of improvisations: 300

are extracted from the z planes of the MR with values of 1, 2, 5, 10, 36, 72, 108 and 144 for de z-axis. Those values are selected in order to acquire representative pictures of different sections of the brain. For MR brain images $nt \in [2, 3, 4, 5]$.

Table 10.2 reports the averaged mean and STD of the fitness function of MCET-CSA. Each experiment for every image and nt consist of 35 independent evaluations of the MCET-CSA. Since it is a minimization problem, the mean is expected to be as low as possible. From the three algorithms implemented (CSA, DE, and HS), CSA is able to find the best threshold values that minimize the objective function.

Table 10.3 presents the averaged results of the quality metrics (PSNR, SSIM, and FSIM) of the segmented images. Each segmented image is generated the threshold values calculated by an ECT. In this case, CSA, DE, and HS are used. From de quantitative analysis MCET-CSA provides evidence of high quality segmented images.

A qualitative comparison of segmented MR brain images is presented in Table 10.4. Four images of the medical benchmark set are selected to visually show the results of the implemented methods. For the sake of visibility, instead of displaying the images on grayscale the colormap is changed for the so-called jet. With this representation, the cold colors indicate low-intensity while the hotter stand for high-intensity values. The images segmented with MCET-CSA provide better contours and a more accurate representation. The results on Table 10.4 were obtained using five thresholds.

10.5.3 Comparative Study

Since the results are originated from similar experiments, the significance of the data must be assessed. For this purpose, Wilcoxon's rank test [25] is conducted with 35 independent samples considering a 5% significance level considering the objective function's value. Table 10.5 presents the p-values generated by Wilcoxon's test for a pair-wise comparison of the fitness value between two groups formed by CSA vs. DE and CSA vs. HS when applied on MR brain images. The results indicate that the values generated by CSA did not occur randomly.

Table 10.2 Results after applying MCET-CSA to the MR brain benchmark images

Img	nt	CSA		DE		HS	
		Mean	STD	Mean	STD	Mean	STD
Z1	2	3.4002	0	3.454	0.0422	3.4209	0.0309
	3	1.6976	0	1.7636	0.0401	1.7201	0.0142
	4	0.9924	0	1.0929	0.061	1.0142	0.0140
	5	0.7055	0.0068	0.8297	0.0692	0.7287	0.0105
Z2	2	3.3614	0	3.4037	0.0296	3.3783	0.0219
	3	1.6694	0	1.7382	0.0408	1.6947	0.0181
	4	0.9882	0	1.0736	0.0472	1.0106	0.0107
	5	0.7054	0.0020	0.825	0.0713	0.7260	0.0090
Z5	2	3.4281	0	3.4595	0.0269	3.4482	0.0206
	3	1.5929	0	1.6708	0.0527	1.6203	0.0142
	4	0.9991	0	1.0856	0.0369	1.0312	0.0469
	5	0.7004	0.0089	0.8103	0.0502	0.7305	0.0150
Z10	2	3.3689	0	3.4244	0.0364	3.4009	0.0307
	3	1.6002	0	1.6981	0.0574	1.6233	0.0149
	4	1.1008	0	1.2047	0.0476	1.1232	0.0287
	5	0.7062	0	0.8439	0.0594	0.7637	0.0463
Z36	2	3.2285	0	3.2796	0.0337	3.2572	0.0280
	3	1.7369	0.0042	1.8092	0.0429	1.7610	0.0131
	4	1.1412	0	1.2263	0.0413	1.1697	0.0311
	5	0.6853	0	0.8386	0.0737	0.7326	0.0227
Z72	2	2.0598	0	2.1231	0.0584	2.0790	0.0262
	3	1.1428	0	1.2526	0.0714	1.1658	0.0220
	4	0.6606	0	0.7989	0.0737	0.7078	0.0290
	5	0.4812	0.0203	0.607	0.0728	0.5131	0.0183
Z108	2	2.0561	0	2.1156	0.0485	2.0797	0.0248
	3	1.1331	0	1.2286	0.053	1.1678	0.0514
	4	0.6292	0	0.7691	0.0877	0.6762	0.0244
	5	0.4879	0.0116	0.6162	0.0589	0.5197	0.0170
Z144	2	1.8075	0	1.8649	0.0354	1.8325	0.0240
	3	1.1050	0	1.1639	0.0239	1.1238	0.0125
	4	0.6919	0.0083	0.7774	0.0459	0.7721	0.0394
	5	0.4128	0	0.5201	0.059	0.4334	0.0114

Table 10.3 Quality evaluation results after applying MCET-CSA to MR brain images where the number of the image title indicate de z plane of the analyzed image

Img	nt	CSA			DE			HS		
		PSNR	SSIM	FSIM	PSNR	SSIM	FSIM	PSNR	SSIM	FSIM
Z1	2	14.2652	0.4962	0.6076	14.1969	0.4938	0.6046	14.1942	0.4937	0.6046
	3	18.0139	0.6771	0.7300	17.8876	0.6728	0.7257	17.9208	0.6737	0.7265
	4	19.8847	0.7377	0.8134	19.7455	0.7322	0.8074	19.7858	0.7340	0.8094
	5	21.5104	0.7950	0.8484	21.2603	0.7836	0.8364	21.6730	0.7897	0.8457
Z2	2	14.3184	0.4940	0.6059	14.2475	0.4916	0.6030	14.2472	0.4915	0.6029
	3	18.0801	0.6773	0.7326	17.9703	0.6743	0.7291	17.9901	0.6739	0.7290
	4	19.9443	0.7366	0.8138	19.7855	0.7300	0.8068	19.8458	0.7329	0.8098
	5	21.7553	0.7922	0.8509	21.3264	0.7820	0.8389	21.7684	0.7878	0.8474
Z5	2	14.6159	0.4968	0.6114	14.5432	0.4943	0.6084	14.5432	0.4943	0.6084
	3	18.2657	0.6803	0.7490	18.1763	0.6758	0.7439	18.1748	0.6769	0.7453
	4	19.9872	0.7305	0.8221	19.8873	0.7262	0.8174	19.8860	0.7268	0.8169
	5	21.4894	0.7894	0.8574	21.2333	0.7819	0.8472	21.8387	0.7811	0.8532
Z10	2	16.2405	0.5860	0.6642	16.1504	0.5830	0.6609	16.1597	0.5831	0.6609
	3	18.6976	0.6701	0.7642	18.5934	0.6656	0.7596	18.6046	0.6668	0.7604
	4	20.2821	0.7085	0.8205	20.1279	0.7023	0.8125	20.1848	0.7050	0.8158
	5	21.6823	0.7854	0.8622	21.7720	0.7768	0.8548	21.7906	0.7711	0.8557
Z36	2	15.9119	0.6338	0.6807	15.8327	0.6306	0.6773	15.8327	0.6306	0.6773
	3	18.4363	0.7149	0.7844	18.2945	0.7099	0.7798	18.1310	0.7065	0.7778
	4	20.1750	0.7559	0.8414	20.0269	0.7480	0.8332	19.9601	0.7530	0.8349
	5	21.2671	0.8209	0.8775	21.0218	0.8082	0.8655	21.1630	0.8168	0.8731

(continued)

Table 10.3 (continued)

Img	nt	CSA			DE			HS		
		PSNR	SSIM	FSIM	PSNR	SSIM	FSIM	PSNR	SSIM	FSIM
Z72	2	15.7358	0.6682	0.7123	15.6544	0.6648	0.7089	15.6575	0.6649	0.7088
	3	19.2135	0.7882	0.8267	19.0740	0.7825	0.8211	19.1179	0.7843	0.8226
	4	20.6608	0.8500	0.8699	20.4755	0.8416	0.861	20.5580	0.8458	0.8656
	5	22.9666	0.8967	0.9042	22.8000	0.8849	0.8942	22.9747	0.8947	0.9009
Z108	2	16.1011	0.6862	0.6913	16.0332	0.6831	0.6881	16.0210	0.6828	0.6879
	3	19.8096	0.8083	0.7900	19.7081	0.8033	0.7844	19.6509	0.8029	0.7848
	4	21.4354	0.8778	0.8538	21.2007	0.8692	0.8441	21.3288	0.8734	0.8496
	5	22.6324	0.8994	0.8823	22.6305	0.8969	0.8746	22.0668	0.8861	0.8717
Z144	2	18.3468	0.7858	0.7616	18.2564	0.7814	0.7575	18.2555	0.7819	0.7578
	3	21.0578	0.8350	0.8331	20.9497	0.8305	0.8287	20.9530	0.8308	0.829
	4	22.5162	0.8968	0.8675	22.2878	0.8871	0.858	22.4511	0.8636	0.8581
	5	24.2984	0.9132	0.9060	24.0175	0.9057	0.8966	24.1757	0.9087	0.9015

Table 10.4 Qualitative results of MCET-CSA applied to MR brain images with $nt = 5$

10.6   Conclusions

This chapter introduces the use of the minimum cross entropy and the crow search algorithm for the segmentation of Magnetic Resonance (MR) brain images. For this purpose, the multilevel thresholding problem is selected due to its high modality nature and its complexity. The proposed methodology considers the thresholding process as an optimization problem, where CSA searches the optimal threshold points considering cross entropy as objective function. The CSA uses particles to encode a set of candidate threshold points. Cross entropy is used as an objective function, evaluating the quality of the selected threshold points. Following the values of the objective function, the operators of CSA guide the evolution process while improving the segmentation of MR brain images. To assess the quality of the seg-

Table 10.5 Wilcoxon p-values of the compared CSA vs. DE and CSA vs. HS on the MR brain image benchmark

Img	Nt	p-values	
		CSA vs. DE	CSA vs. HS
Z1	2	6.82E−13	2.35E−12
	3	9.19E−13	3.09E−09
	4	7.11E−13	1.22E−08
	5	1.29E−12	2.35E−05
Z2	2	4.27E−13	4.99E−12
	3	1.18E−12	8.87E−09
	4	1.14E−12	9.24E−09
	5	1.41E−12	8.49E−04
Z5	2	4.22E−13	7.11E−13
	3	1.29E−12	1.01E−09
	4	7.74E−13	9.64E−08
	5	8.82E−13	1.69E−06
Z10	2	5.01E−13	1.14E−12
	3	9.96E−13	5.24E−09
	4	6.53E−13	2.68E−05
	5	6.90E−12	7.67E−03
Z36	2	4.91E−13	1.97E−11
	3	3.00E−12	5.54E−08
	4	1.19E−12	7.17E−04
	5	4.38E−12	1.14E−04
Z72	2	5.49E−13	1.84E−10
	3	1.41E−12	9.97E−08
	4	7.13E−13	1.15E−05
	5	2.55E−12	1.39E−02
Z108	2	6.63E−13	2.81E−12
	3	7.11E−13	1.68E−09
	4	6.53E−13	2.77E−06
	5	2.77E−12	5.53E−04
Z144	2	5.30E−13	4.32E−11
	3	7.76E−13	1.45E−08
	4	7.48E−12	8.70E−10
	5	1.41E−12	2.98E−02

mented images, STD, PSNR, SSIM, and FSIM are used. Those metrics consider the similarity between the original and the segmented images.

For comparisons two Evolutionary Computation Techniques are implemented, the Differential Evolution (DE) and Harmony Search (HS). The competence of the algorithms is assessed regarding PSNR, STD, SSIM, FISM and fitness values. Such comparisons evidence the convergence, accuracy and robustness of the CSA, in contrast with those of the DE and HS. Wilcoxon's test is used to determinate that the results of the CSA are significantly different from the ones of DE and HS and did not occur by chance. Quantitative results indicate that MCET-CSA generates high quality segmented MRI regarding PSNR, SSIM, and FSIM. Besides, the qualitative analysis of the results after segmenting brain images on different depths shows well-delimited regions that are easier to distinguish in comparison to other techniques.

References

1. Lee S, Jang J, Baek K-R, Shim H (2014) Modified Otsu's method for indoor mobile robot tracking system. In: 2014 International Conference on Electronics, Information and Communications (ICEIC). IEEE, pp 1–2
2. Jia C, Qi J, Li X, Lu H (2016) Saliency detection via a unified generative and discriminative model. Neurocomputing 173:406–417. https://doi.org/10.1016/j.neucom.2015.03.122
3. Ji Z, Liu J, Cao G et al (2014) Robust spatially constrained fuzzy c-means algorithm for brain MR image segmentation. Pattern Recognit 47:2454–2466. https://doi.org/10.1016/j.patcog.2014.01.017
4. Hammouche K, Diaf M, Siarry P (2010) A comparative study of various meta-heuristic techniques applied to the multilevel thresholding problem. Eng Appl Artif Intell 23:676–688. https://doi.org/10.1016/j.engappai.2009.09.011
5. Akay B (2013) A study on particle swarm optimization and artificial bee colony algorithms for multilevel thresholding. Appl Soft Comput J 13:3066–3091. https://doi.org/10.1016/j.asoc.2012.03.072
6. Liao PS, Chen TS, Chung PC (2001) A fast algorithm for multilevel thresholding. J Inf Sci Eng 17:713–727
7. Otsu N (1979) Threshold selection method from gray-level histograms. IEEE Trans Syst Man Cybern SMC 9:62–66
8. Kapur JN, Sahoo PK, Wong AKC (1985) A new method for gray-level picture thresholding using the entropy of the histogram. Comput Vis Graph Image Process 29:273–285
9. Kittler J, Illingworth J (1986) Minimum error thresholding. Pattern Recognit 19:41–47. https://doi.org/10.1016/0031-3203(86)90030-0
10. Sankur B (2004) Survey over image thresholding techniques and quantitative performance evaluation. J Electron Imaging 13:146. https://doi.org/10.1117/1.1631315
11. de Albuquerque MP, Esquef IA, Gesualdi Mello AR (2004) Image thresholding using Tsallis entropy. Pattern Recognit Lett 25:1059–1065. https://doi.org/10.1016/j.patrec.2004.03.003
12. Li CH, Tam PKS (1998) An iterative algorithm for minimum cross entropy thresholding. Pattern Recognit Lett 19:771–776. https://doi.org/10.1016/S0167-8655(98)00057-9
13. Li CH, Lee CK (1993) Minimum cross entropy thresholding, Pattern recognition, vol 26(4), pp 617–625, https://doi.org/10.1016/0031-3203(93)90115-D
14. Yin P-YP (2007) Multilevel minimum cross entropy threshold selection based on particle swarm optimization. Appl Math Comput 184:503–513. https://doi.org/10.1109/SNPD.2007.85
15. Horng M-H, Liou R-J (2011) Multilevel minimum cross entropy threshold selection based on the firefly algorithm. Expert Syst Appl 38:14805–14811. https://doi.org/10.1016/j.eswa.2011.05.069

16. Oliva D, Osuna-Enciso V, Cuevas E et al (2015) Improving segmentation velocity using an evolutionary method. Expert Syst Appl 42:5874–5886. https://doi.org/10.1016/j.eswa.2015.03.028
17. Storn R, Price K (1997) Differential evolution – a simple and efficient heuristic for global optimization over continuous spaces. J Glob Optim 11:341–359. https://doi.org/10.1023/A:1008202821328
18. Loganathan GV (2001) A new heuristic optimization algorithm: harmony search. Simulation 76:60–68. https://doi.org/10.1177/003754970107600201
19. Kullback S (1968) Information theory and statistics (Dover books on mathematics). Dover, Mineola
20. Askarzadeh A (2016) A novel metaheuristic method for solving constrained engineering optimization problems: crow search algorithm. Comput Struct 169:1–12. https://doi.org/10.1016/j.compstruc.2016.03.001
21. Black H (2013) Social skills to crow about. Sci Am Mind 24:12. https://doi.org/10.1038/scientificamericanmind0913-12
22. Geem ZW, Kim JH, Loganathan GV (2001) A new heuristic optimization algorithm: harmony search. Simulation 762:60–68
23. Pedersen M, Good parameters for particle swarm optimization. hvass-labs.org
24. Oliva D, Cuevas E, Pajares G, et al (2013) Multilevel thresholding segmentation based on harmony search optimization, J Appl Math 2013(575414):24. https://doi.org/10.1155/2013/575414
25. García S, Molina D, Lozano M, Herrera F (2008) A study on the use of non-parametric tests for analyzing the evolutionary algorithms' behaviour: a case study on the CEC' 2005 special session on real parameter optimization. J Heuristics 15:617–644. https://doi.org/10.1007/s10732-008-9080-4

Chapter 11
Fuzzy Entropy Approaches for Image Segmentation

11.1 Introduction

In previous chapters they were analyzed some of the most important entropy-based techniques for image thresholding. Entropy is a concept firstly used in the second law of Thermodynamics; it was introduced into physics by German physicist Rudolf Clausius in second half of 18th century. In general terms, in the image processing field it is possible to said that the Kapur entropy [1] is the most popular approach used for finding the best thresholds that segment a digital image. Kapur proposes the maximization of the entropy as a measure of the homogeneity among classes. The Kapur entropy has been used for different implementations, for example for the segmentation of thermal images [2], breast cancer images [3] and breast histology images [4]. However, the use of Kapur entropy doesn't guarantee to obtain the best results in complex images. This fact occurs when the number of thresholds increase because each threshold increases the complexity to compute the entropy and decrease the accuracy of the segmentation.

Different entropies have been proposed to be used instead Kapur, some examples are the Shannon entropy [5], Tsallis entropy [6], Rényi's entropy [7, 8], Cross entropy [9] and the generalized entropy [10]. In Information theory, Shannon's entropy measures the information contained in a dataset, but this information is not the meaningful. Here is important to mention that Shannon is the base of all other entropies used for image segmentation. The Rényi's entropy is a modified version of the Shannon entropy that includes the maximum entropy sum method, and the entropic correlation method. The combination of entropies presented by Rényi provides better results that using them separately. Tsallis and Rényi entropy measures are two possible different generalizations of the Shannon's entropy but are not generalizations of each other. Considering the above it has been proposed a generalized entropy introduced by Masi [10]. This method has a parameter that measures the degree of additivity and/or non-extensivity of the data. Such value must be properly tuned to obtain the best segmentation results. There are several entropies used for image segmentation,

© Springer Nature Switzerland AG 2019   141
D. Oliva et al., *Metaheuristic Algorithms for Image Segmentation:
Theory and Applications*, Studies in Computational Intelligence 825,
https://doi.org/10.1007/978-3-030-12931-6_11

all of them produce different results and the accuracy also vary among them. Moreover, complex images have overlapped classes that must be accurately separated.

On the other hand, the concept of Fuzzy entropy has been introduced in [11] as an alternative to classical entropy segmentation approaches. In general terms the fuzzy based approaches remove the greyness ambiguities presented in the image, this fact is reflected on the accuracy of the segmented images. The concept of fuzzy entropy was firstly proposed by De Luca et al. in 1972 [12]. This kind on entropy uses a function that works with fuzzy sets, its value becomes smaller when the sharpness of its argument fuzzy set is improved. In recent years, numerous image segmentation methods based on fuzzy entropy have been proposed. However, in this chapter are examined only the classical fuzzy entropy and the type II fuzzy entropy.

This chapter is organized as follows: Sect. 11.2 describes the main concepts of fuzzy entropy and type II fuzzy entropy. Meanwhile, Sect. 11.3 presents a generic formulation of the fuzzy entropy as an optimization problem. Finally, Sect. 11.4 discuss some conclusions.

11.2   Fuzzy Entropy

Images contain different ambiguities generated by the scenes, for that reason is necessary to use robust image processing methods. The fuzzy set theory is mathematical tool commonly used to handle uncertainty and ambiguity. In this sense the fuzzy entropy (FE) is used for image thresholding, the methodology used considers the fuzzy memberships as an indication of how strongly a gray value belongs to the background or to the foreground. This section introduces the concepts of FE and the Type II FE (TII-FE).

11.2.1   The Concepts of Fuzzy Entropy

The Fuzzy Entropy (also known as Type I Fuzzy Entropy) works with the histogram (h) of a digital image, it also assumed that the total amount of gray levels is $L = 256$. There is necessary to obtain probability of a gray level i that is computed as $p(i) = h(i)/N$, N is the total number of pixels contained in the image. As was explained in previous chapters for multilevel thresholding it is necessary to use a set of threshold values defined as $\mathbf{TH} = [th_1, th_2, \ldots, th_n]$ to divide the histogram into $n + 1$ classes. These assumptions are also considered for the FE but it also uses a trapezoidal membership function (Eq. 11.1) to estimate the degree of inclusion of the $k = n + 1$ classes.

$$\mu_k(i) = \begin{cases} 0 & \text{if} \quad i \le a_{k-1} \\ \frac{i - a_{k-1}}{c_{k-1} - a_{k-1}} & \text{if} \quad a_{k-1} < i \le c_{k-1} \\ 1 & \text{if} \quad c_{k-1} < i \le a_k \\ \frac{i - c_k}{a_k - c_k} & \text{if} \quad a_k < i \le c_k \\ 0 & \text{if} \quad i > c_k \end{cases} \qquad (11.1)$$

From Eq. (11.1) it is necessary to define the best values for the fuzzy parameters a_k and c_k of the membership function. To estimate the values of such parameters the following rule must be considerated: $0 \le a_1 \le c_1 \le a_2 \le c_2 \le \cdots \le a_n \le c_n \le L - 1$. Using the values obtained by Eq. (11.1) the fuzzy entropy of an image is computed as:

$$H = H_1 + H_2 + \cdots + H_{n+1} = -\sum_{i=0}^{L-1} \left(\frac{p_i * \mu_1(i)}{P_1} \right) * \ln \sum_{i=0}^{L-1} \left(\frac{p_i * \mu_1(i)}{P_1} \right)$$

$$-\sum_{i=0}^{L-1} \left(\frac{p_i * \mu_2(i)}{P_2} \right) * \ln \sum_{i=0}^{L-1} \left(\frac{p_i * \mu_2(i)}{P_2} \right)$$

$$-\cdots -\sum_{i=0}^{L-1} \left(\frac{p_i * \mu_{n+1}(i)}{P_{n+1}} \right) * \ln \sum_{i=0}^{L-1} \left(\frac{p_i * \mu_{n+1}(i)}{P_{n+1}} \right) \qquad (11.2)$$

In Eq. (11.2) the values of P are computed as follows:

$$P_1 = \sum_{i=1}^{th_1} p_i \mu_1(i), \quad P_2 = \sum_{i=th_1+1}^{th_2} p_i \mu_2(i), \ldots, P_{n+1} = \sum_{i=th_n+1}^{L-1} p_i \mu_{n+1}(i), \quad (11.3)$$

Considering Eqs. 11.2 and 11.3 the optimal fuzzy parameters are obtained as:

$$(a_1, c_1, a_2, c_2, \ldots, a_n, c_n) = \max(H) \qquad (11.3)$$

The best thresholds used to segment the image are extracted using the fuzzy parameters as in Eq. (11.4).

$$th_1 = \frac{1}{2}(a_1 + c_1), \quad th_2 = \frac{1}{2}(a_2 + c_2), \ldots, th_n = \frac{1}{2}(a_n + c_n) \qquad (11.4)$$

11.2.2 Type II Fuzzy Entropy

The Type II Fuzzy sets (TII-FS) are a generalization of Type I Fuzzy sets. In TII-FS is introduced the concept of ultrafuzziness that determine and eliminate the uncertainties that the classical Fuzzy sets possess [13]. According to the related literature, the Type II Fuzzy Entropy (TII-FE) is one of the easiest and faster techniques based on TII-FS [14, 15]. However, similar to the type I FE the main drawback of TII-FE is that it requires estimating the fuzzy parameters of the membership functions. This problem becomes worst in determining the number of thresholds [16], this fact is reflected in the computational effort required to compute the entropy.

Since the TII-FE is an extension of the FE, it also works with the histogram and with the same considerations. In this sense, in this section there are introduced the features of TII-FE, in specific the ultra-fuzziness. The ultra-fuzziness can be used as a metric associated to a fuzzy set which gives a 0 value when the membership values can be represented without any uncertainty. Whereas the value rises to 1 when membership values can be specified within an interval. For a digital image, the ultra-fuzziness for the i-th level of intensity is defined as follows:

$$P_k = \sum_{i=0}^{L-1} \left(h_i * \left(\mu_k^{high}(i) - \mu_k^{low}(i) \right) \right), \ k = \{1, 2, \ldots, nl\}, \tag{11.5}$$

where μ_k is the trapezoidal fuzzy membership function that generates the sets where the pixels belong depending on their intensity value. The Eq. (11.6) provides a better explanation of the membership function used in this approach.

$$\mu_k(i) = \begin{cases} 0 & \text{if} \quad i \leq a_{k-1} \\ \frac{i-a_{k-1}}{c_{k-1}-a_{k-1}} & \text{if} \ a_{k-1} < i \leq c_{k-1} \\ 1 & \text{if} \ c_{k-1} < i \leq a_k \\ \frac{i-c_k}{a_k-c_k} & \text{if} \ a_k < i \leq c_k \\ 0 & \text{if} \quad i > c_k \end{cases} \tag{11.6}$$

In Eq. (11.6) a_k and c_k are depicted as the fuzzy parameters where $k = \{1, 2, \ldots, nl\}$. The values for the bounds are given as: $a_0 = c_0 = 0$ and $a_{nl+1} = c_{nl+1} = nl + 1$, where nl denotes the number of thresholds used for segmentation. The fuzzy type-II entropy for a k-th threshold th_k is therefore given as:

$$Fe_k = -\sum_{i=1}^{L-1} \left(\frac{\left(h_i * \left(\mu_k^{high}(i) - \mu_k^{low}(i) \right) \right)}{P_k} \right) * \ln \left(\frac{\left(h_i * \left(\mu_k^{high}(i) - \mu_k^{low}(i) \right) \right)}{P_k} \right), \ k = \{1, 2, \ldots, nl + 1\} \tag{11.7}$$

The sum of all the entropies for the $nl + 1$ levels is the total entropy defined as:

$$T_{Fe}(a_1, c_1, \ldots, a_n, c_n) = \sum_{k=1}^{nt+1} Fe_i \tag{11.8}$$

The problem of Eq. (11.8) is to define the best values for the fuzzy parameters, this process can be performed if the total entropy is maximized as in Eq. (11.9).

$$(a_1, c_1, \ldots, a_n, c_n) = \max(T_{Fe}) \tag{11.9}$$

Finally, the best thresholds are obtained using the Eq. (11.4).

11.3 Fuzzy Entropy as an Optimization Problem

This section explains how the fuzzy entropy approaches can be treated as an optimization problem. The fuzzy entropy is different to other entropies, for that reason it is necessary to define how to formulate the solutions. For example, for Kapur's entropy the problem is to find the best thresholds that provide the highest entropy. In this context when is used an optimization algorithm for searching the best thresholds, the method considers each candidate solution as a set of values that optimize the entropy function. Such sets are conformed by thresholds and are modified according with the operators of the optimization algorithm.

On the other hand, for the fuzzy entropy it is necessary to estimate the parameters that define the membership functions a_k and c_k. Each solution in the set then contains the double amount of thresholds $2 \times nt$. According with the definition of the FE and the TII-FE it is necessary to define two dummy thresholds 0 and $L - 1$ such values are also included in the candidate solutions. In Eq. (11.10) is suggested the construction of a candidate fuzzy solution and the entire set of solutions.

$$\mathbf{G} = [\mathbf{Gf}_1, \mathbf{Gf}_2, \ldots, \mathbf{Gf}_N],$$
$$\mathbf{Gf}_i = [0, 0, a_1, c_1, a_2, c_2, \ldots, a_n, c_n, L - 1, L - 1]^T \qquad (11.10)$$
$$\mathbf{Gf}_i \subseteq \mathbf{G}$$

Once is defined the formulation of the solutions any optimization algorithm can be used (swarm, evolutionary, physically inspired, etc.). However, it is important to mention that for the initialization of the candidate solution the bounds of the search space are stablished by the histogram. In this sense, the lower bound is zero and the upper bound is 255. Such bounds are extended for all the dissension of the problem.

No matter which optimization method is selected to search the best fuzzy parameters, the important thing to define if it used the fuzzy entropy or the type II fuzzy entropy. The designer then needs to select the objective function that permits to evaluate the quality of the candidate solutions. The objective function where defined in the Sect. 11.2, for fuzzy entropy the reader must select Eq. (11.3), meanwhile for TII-FE it must select Eq. (11.9).

11.4 Summary

This chapter introduces the theory that permits the implementation of metaheuristic algorithms for image segmentation using fuzzy entropy and type II fuzzy entropy. Moreover, here is also presented the way that the solutions must be created and in order to be modified by any optimization approach.

This kind of entropies are interesting alternatives to other classical methods. They permit to handle complex images that presents ambiguities and uncertainties that are not easily visible. Since the aim of this chapter is only presents the basic concepts of fuzzy entropy approaches for images thresholding, we encourage the reader to implement them using the datasets used in other chapters. The reader can use some standard algorithms like the Particle Swarm Optimization (PSO) [17], Differential Evolution (DE) [18] or any other modern approach. Some comparison can be performed with the results presented in this book for other entropy approaches. Then the reader can analytically decide which entropy is better for a specific implementation.

References

1. Kapur JN, Sahoo PK, Wong AKC (1985) A new method for gray-level picture thresholding using the entropy of the histogram. Comput Vis Graph Image Process 29:273–285
2. Hinojosa S, Pajares G, Cuevas E, Ortega-Sanchez N (2018) Thermal image segmentation using evolutionary computation techniques. Studies in computational intelligence, vol 730, pp 63–88. https://doi.org/10.1007/978-3-319-63754-9_4
3. Díaz-Cortés MA, Ortega-Sánchez N, Hinojosa S et al (2018) A multi-level thresholding method for breast thermograms analysis using Dragonfly algorithm. Infrared Phys Technol 93:346–361. https://doi.org/10.1016/j.infrared.2018.08.007
4. Hinojosa S, Dhal KG, Elaziz MA et al (2018) Entropy-based imagery segmentation for breast histology using the Stochastic Fractal Search. Neurocomputing 321:201–215. https://doi.org/10.1016/j.neucom.2018.09.034
5. Benzid R, Arar D, Bentoumi M (2008) A fast technique for gray level image thresholding and quantization based on the entropy maximization. In: 5th international multi-conference system signals devices. IEEE, Amman, pp 1–4
6. Sarkar S, Das S, Chaudhuri SS (2012) Multilevel image thresholding based on Tsallis entropy and differential evolution. Swarm, evolutionary, and memetic computing, SEMCCO 2012. https://doi.org/10.1007/978-3-642-35380-2_3
7. Sahoo PK, Arora G (2004) A thresholding method based on two-dimensional Renyi's entropy. Pattern Recognit 37:1149–1161. https://doi.org/10.1016/j.patcog.2003.10.008
8. Liu SL, Kong LZ, Wang JG (2010) Segmentation approach based on fuzzy Renyi entropy
9. Pal NR (1996) On minimum cross-entropy thresholding. Pattern Recognit 29:575–580. https://doi.org/10.1016/0031-3203(95)00111-5
10. Masi M (2005) A step beyond Tsallis and Rényi entropies. Phys Lett Sect A Gen At Solid State Phys 338:217–224. https://doi.org/10.1016/j.physleta.2005.01.094
11. Tian W, Geng Y, Liu J, Ai L (2009) Maximum fuzzy entropy and immune clone selection algorithm for image segmentation. In: 2009 Asia-Pacific Conference Information Processing. IEEE, Shenzhen, pp 38–41
12. De Luca A, Termini S (1972) A definition of a nonprobabilistic entropy in the setting of fuzzy sets theory. Inf Control 20:301–312. https://doi.org/10.1016/S0019-9958(72)90199-4
13. Tizhoosh HR (2008) Type II fuzzy image segmentation. Studies in fuzziness and soft computing, vol 220, pp 607–619. https://doi.org/10.1007/978-3-540-73723-0_31
14. Tizhoosh HR (1998) Fuzzy image processing (in German). https://doi.org/10.1007/978-3-642-58742-9
15. Tizhoosh HR (2005) Image thresholding using type II fuzzy sets. Pattern Recognit 38:2363–2372. https://doi.org/10.1016/j.patcog.2005.02.014
16. Burman R, Paul S, Das S (2013) A differential evolution approach to multi-level image thresholding using type II fuzzy sets. Lecture notes in computer science (including subseries lecture notes in artificial intelligence and lecture notes in bioinformatics), LNCS, vol 8297, pp 274–285. https://doi.org/10.1007/978-3-319-03753-0_25

17. Kennedy J, Eberhart RC (1995) Particle swarm optimization. In: 1995 proceedings of the IEEE international conference on neural networks, vol 4, pp 1942–1948. https://doi.org/10.1109/icnn.1995.488968
18. Storn R, Price K (1997) Differential evolution - a simple and efficient heuristic for global optimization over continuous spaces. J Glob Optim 11:341–359

Chapter 12
Image Segmentation by Gaussian Mixture

12.1 Introduction

Until now all the methods described in this book are based on entropy or variance functions. They are considered as no parametric approaches since for the segmentation is only necessary to find the thresholds for the histogram. However, these techniques do not consider the shape of the histogram. One of the most used proposals used to approximate the histogram is the gaussian mixture that defines the histogram as a combination of probability density function defined by the Gaussian distribution. The parameters used in the combination of functions needs to be properly selected. The estimation of such parameters is considered as a non-linear optimization problem. The unknown parameters that give the best fit to the processed histogram are determined by using the so-called differential evolution optimization algorithm.

The aim of this chapter is to introduce the main concepts and the formulation of the optimization problem using the gaussian mixture approach. Moreover, here is also explained the process of extract the thresholds that segment the histogram of a digital image.

12.2 Theory of Gaussian Approximation

The gaussian approximation also works with the image histogram $h(p)$ that represents the distribution of the gray levels L of a digital image. The histogram must be normalized to be considered as a probability distribution function according to Eq. (12.1).

$$h(p) = \frac{n_p}{N}, \ h(p) \geq 0, \ N = \sum_{p=0}^{L-1} n_p, \ \text{and} \ \sum_{p=0}^{L-1} h(p) = 1 \qquad (12.1)$$

© Springer Nature Switzerland AG 2019   149
D. Oliva et al., *Metaheuristic Algorithms for Image Segmentation:
Theory and Applications*, Studies in Computational Intelligence 825,
https://doi.org/10.1007/978-3-030-12931-6_12

n_p corresponds to the number of pixels with gray level p, N defines the total number of pixels in the image. On the other hand, the histogram function can be defined as a mixture of gaussian function as follows:

$$b(x) = \sum_{i=1}^{K} B_i \cdot b_i(x) = \sum_{i=1}^{K} \frac{B_i}{\sqrt{2\pi}\sigma_i} \exp\left[\frac{-(x - \mu_i)^2}{2\sigma_i^2}\right] \qquad (12.2)$$

From Eq. (12.2) B_i is the a priori probability of class i, $b_i(x)$ is the probability function of a random gray level value x in the class i. Meanwhile, μ_i and σ_i are the mean and the standard deviation of the i-th probability distribution. Finally, K corresponds to the number of classes in the image. In Eq. (12.2) there is necessary to estimate the parameters B_i, μ_i and σ_i, for $i = 1, 2 \dots, K$. This problem also includes the following constraint that must be satisfied:

$$\sum_{i=1}^{K} P_i = 1 \qquad (12.3)$$

To estimate the $3K$ parameters it is used the mean square error that exists between the composite gaussian function $b(x_i)$ and the histogram $h(x_i)$ and it is defined as:

$$E = \frac{1}{n} \sum_{j=1}^{n} \left(b(x_j) - h(x_j)\right)^2 + \omega \cdot \left|\left(\sum_{i=1}^{K} B_i\right) - 1\right| \qquad (12.4)$$

From [13 book] we can assume a point n in the histogram and ω is a penalty factor for the constraint defined in Eq. (12.3).

The estimation of the best parameters that minimize the square error is a complex task since image segmentation is a crucial step in many applications [1].

An easy method for search the optimal parameters decreases the partial derivatives of Eq. (12.4) to zero but it considers transcendental equations [2]. In this sense, due to the non-linearity of such equations it is not possible to obtain an analytical solution. This algorithm also employs gradient operations that in many cases get stuck within local minima and the initial configuration of the solution also affects the final solution.

In Fig. 12.1 is presented an example of the gaussian mixture presented in this section. Figure 12.1(a) shows a histogram that at this moment doesn't represent any image, for illustration it has 3 classes. Figure 12.1(b) shows the 3 gaussian functions that are computed using Eq. (12.2). In the same context the Fig. 12.1(c) demonstrate (in red) the gaussian mixture.

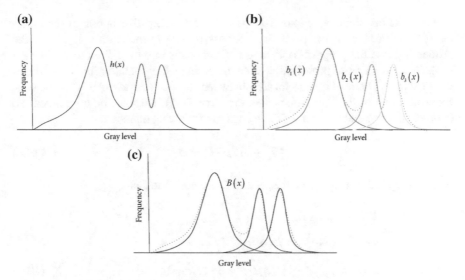

Fig. 12.1 An example of the gaussian mixture, **a** represents the original histogram, **b** shows the gaussians computed and **c** presents the gaussian mixture

12.3 Extraction of the Threshold Values

As was previously mentioned the thresholding is the easiest method for image segmentation. This section explains how to obtain the thresholds once the best histogram approximation is generated by the parameters of the gaussian mixture and the objective function.

The classes in the approximated histogram are sorted as $\mu_1 < \mu_2 < \ldots < \mu_K$. Considering the above the thresholds are generated by the calculation of the total probability error for two adjacent gaussian functions. This process is explained in Eqs. (12.5)–(12.7).

$$E(T_i) = B_{i+1} \cdot E_1(T_i) + B_i \cdot E_2(T_i), \ i = 1, 2, \ldots, K - 1 \tag{12.5}$$

where:

$$E_1(T_i) = \int_{-\infty}^{T_i} b_{i+1}(x)dx \tag{12.6}$$

and

$$E_2(T_i) = \int_{T_i}^{\infty} b_i(x)dx \tag{12.7}$$

From the last three equations $E_1(T_i)$ is the probability distribution of the incorrectly classification of the pixels in the class $(i + 1)$ to the class i. $E_2(T_i)$ is the probability of a not proper classification of the pixels into the i class to the class $(i + 1)$. B_i are the a priori probabilities with the combined probability function. Meanwhile, T_i is a threshold value located between the i-th and the $(i + 1)$-th classes. Moreover, one of the T_i is selected as the error $E(T_i)$ that must be minimized. To compute the optimal threshold T_i it is used the following equation:

$$AT_i^2 + BT_i + C = 0 \tag{12.8}$$

From Eq. (12.8) the values of A, B and C are computed as:

$$A = \sigma_i^2 - \sigma_{i+1}^2$$
$$B = 2 \cdot \left(\mu_i \sigma_{i+1}^2 - \mu_{i+1} \sigma_i^2 \right)$$
$$C = (\sigma_i \mu_{i+1})^2 - (\sigma_{i+1} \mu_i)^2 + 2 \cdot (\sigma_i \sigma_{i+1})^2 \cdot \ln\left(\frac{\sigma_{i+1} B_i}{\sigma_i B_{i+1}} \right) \tag{12.9}$$

The quadratic equation give us two possible solutions, however for this problem, only one, the positive is the feasible one.

12.3.1 Solutions Representation

As was explained in the previous section for gaussian mixture it is necessary to accurately define the best parameters. However, it is not an easy task and the complexity increases with the number of classes. To overcome this problem, they can be used optimization algorithms that optimize the Eq. (12.5). In the related literature they have been proposed some interesting implementations [3–5].

This section then explains an alternative to construct the solutions for a standard optimization algorithm. The reader can apply the methodology presented here with any algorithm just considering the order of the matrixes or vectors. In general terms, each gaussian has three parameters (B, σ, μ) then each solution has $3 \times K$ elements, where K is the number of classes to segment. In Eq. (12.10) it is presented the proposed construction of a solution.

$$\mathbf{S} = [\mathbf{s}_1, \mathbf{s}_2, \ldots, \mathbf{s}_N]$$
$$\mathbf{s}_N = \left[B_1^N, \sigma_1^N, \mu_1^N, B_2^N, \sigma_2^N, \mu_2^N, \ldots, B_K^N, \sigma_K^N, \mu_K^N \right] \tag{12.10}$$

In most of the optimization algorithms the first step consists in randomly initialize the set of candidate solutions. They are modified by the operator of each optimization algorithm and then the best solution is extracted at the end of the iterative process.

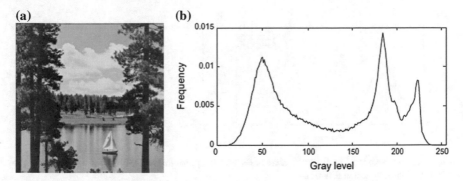

Fig. 12.2 **a** Gray level benchmark image and **b** the corresponding histogram

Finally, the thresholds are computed and applied to the original image according with the procedure explained in the previous chapters.

12.3.2 An Example of Image Segmentation

This section presents an illustrative example of image segmentation using the gaussian mixture model. In Fig. 12.2 is presented a gray level image and its histogram.

For experimental purposes there are considered three classes for the segmentation. In this context each solution has nine elements that must be estimated. Equation (12.11) shows how each solution is constructed.

$$\mathbf{s}_N = \left[B_1^N, \sigma_1^N, \mu_1^N, B_2^N, \sigma_2^N, \mu_2^N, B_3^N, \sigma_3^N, \mu_3^N \right] \tag{12.11}$$

An optimization algorithm like Particle Swarm Optimization (PSO) [6], Differential Evolution (DE) [7] or any other modern approach can be used for searching the best configuration using the objective function introduced in Eq. (12.5). Once the stop criterion is satisfied, the best solution of the set is extracted, and the thresholds are computed using Eq. (12.9). In Fig. 12.3a are presented the three gaussians found by the optimization process. Meanwhile, Fig. 12.3b shows the combination of the three gaussian functions.

From Fig. 12.3(b) the reader can observe that shape is very similar to the histogram of Fig. 12.2(b). It means that the algorithm properly finds the best configuration that reduce the error of the objective function. Finally, in Fig. 12.4 is presented the segmented image using the best thresholds computed using the mixture of gaussian functions.

From Fig. 12.4 all the objects in the scene are properly segmented in a class created by the thresholds computed by approximation of the histogram. We would like to encourage the reader to test the model presented in this chapter with different optimization algorithms. Moreover, the results can be compared with other approaches in

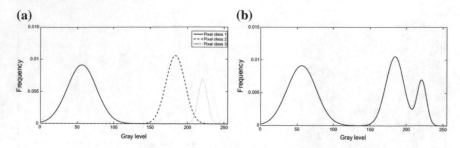

Fig. 12.3 **a** Gaussian functions of each class and **b** approximation of the histogram using the combination of gaussian functions

Fig. 12.4 Segmented image by the gaussian mixture approach

terms of segmentation quality using metrics like feature similarity index (FSIM) [8], the structure similarity index (SSIM) [9] or the peak signal noise ratio (PSNR) [10].

12.4   Summary

In this chapter are presented the concepts that permits to segment an image using a mixture of gaussian function. It is a parametric approach that approximates the histogram using a combination of gaussian distribution. Each class of pixels is formed by a gaussian function. To compute each class, they are necessary three parameters that needs to be computed in order to have a good approximation. This is the main problem of this method because the number of parameters to find increases with the number of classes. In this sense, this methodology can be addressed from an optimization point of view, that is the aim of this chapter.

The mixture of gaussians is an interesting alternative to the non-parametric segmentation approaches (like entropies of variances). It has some properties that can be explored, for example once the histogram is approximated, they can be used other methodologies to extract the thresholds. Here is introduced an equation that permits to compute them easily, but the reader can improve this part of the approach. In addition, some comparison can be performed with the results presented in this book for other entropy approaches. Then the reader can analytically decide which method is better for a specific implementation.

References

1. Hinojosa S, Dhal KG, Elaziz MA, Oliva D, Cuevas E, Entropy-based imagery segmentation for breast histology using the stochastic fractal search. Neurocomputing 321: 201–215. https://doi.org/10.1016/j.neucom.2018.09.034
2. Gonzalez RC, Woods RE (1992) Digital image processing. Pearson, Prentice-Hall, New Jersey
3. Cuevas E, Zaldivar D, Pérez-Cisneros M (2010) A novel multi-threshold segmentation approach based on differential evolution optimization. Expert Syst Appl 37:5265–5271. https://doi.org/10.1016/j.eswa.2010.01.013
4. Cuevas E, González A, Cuevas E et al (2015) An optimisation algorithm based on the behaviour of locust swarms locust swarms. https://doi.org/10.1504/ijbic.2015.073178
5. Osuna-Enciso V, Cuevas E, Sossa H (2013) A comparison of nature inspired algorithms for multi-threshold image segmentation. Expert Syst Appl 40:1213–1219. https://doi.org/10.1016/j.eswa.2012.08.017
6. Kennedy J, Eberhart RC (1995) Particle swarm optimization. In: Proceedings of international conference on neural networks, vol 4, pp 1942–1948. https://doi.org/10.1109/icnn.1995.488968
7. Storn R, Price K (1997) Differential evolution – a simple and efficient heuristic for global optimization over continuous spaces. J Glob Optim 11:341–359
8. Zhang L, Zhang L, Mou X, Zhang D (2011) FSIM: a feature similarity index for image quality assessment. IEEE Trans Image Process 20:2378–2386. https://doi.org/10.1109/TIP.2011.2109730
9. Hinojosa S, Pajares G, Cuevas E, Ortega-Sanchez N (2018) Thermal image segmentation using evolutionary computation techniques. Stud Comput Intell 730:63–88. https://doi.org/10.1007/978-3-319-63754-9_4
10. Suresh S, Lal S (2017) Multilevel thresholding based on Chaotic Darwinian particle swarm optimization for segmentation of satellite images. Appl Soft Comput J 55:503–522. https://doi.org/10.1016/j.asoc.2017.02.005

Chapter 13
Image Segmentation as a Multiobjective Optimization Problem

13.1 Introduction

As was mentioned in previous chapters, image processing techniques have been attracting a considerable amount of attention of technicians, scientists and practitioners due they are applied in different areas in our daily life [1, 2]. For example, medical images which used to provide more information that can be used in diagnostics system [3] to improve the health of human. In Agriculture process [4], the image process plays a critical rule in different aspect starts from the analysis the seed to care the plant and this leads to increase the production, as well as, the economics. For [5], the monitoring the production during the manufacturing process, also, analysis the images taken from cameras that placed at a position where no person can access such as too high (also two low) temperatures in the environments. For more applications, see [6].

Based on all these applications of image processing, they still required preprocessing methods, and the image segmentation is considered one of the most popular preprocessing. It is aimed to group the pixels in different classes based on several criteria. There are several image segmentation methods have been presented such as clustering [7], edge detection [8], thresholding [9]. However, from all of these methods, the thresholding is more attractive than others [10].

In general, the thresholding methods have two main categories of approaches namely bi-level (BL) and multilevel (ML). In the BL, there is only one threshold values that used to split the image into its background and foreground. These methods are suitable only for the image that contains only two classes; however, in the case, the image has more than two classes then these BL methods become unsuitable and time-consuming. To avoid these limitations of BL methods, the ML thresholding methods have been used.

There are several ML thresholding approaches have been proposed to find the optimal threshold values, these methods depend on maximizing/minimizing the objective function. The traditional ML methods provide an acceptable result in the case a small

© Springer Nature Switzerland AG 2019

D. Oliva et al., *Metaheuristic Algorithms for Image Segmentation:*
Theory and Applications, Studies in Computational Intelligence 825,
https://doi.org/10.1007/978-3-030-12931-6_13

number of thresholds is required to segment the image. However, when the number of thresholds increases, it becomes unpractical to use these traditional methods. Therefore, there is another trend to use global optimization methods to find the threshold values such as the meta-heuristic methods. These MH approaches include Particle Swarm Optimization (PSO) [11], Whale optimization algorithm (WOA) [12], Genetic Algorithm (GA) [13], and Salp swarm Algorithm (SSA) [14].

Most of these MH methods that proposed for image segmentation are used Otsu [15] and Kapur [16] as the objective function. Where Otsu aimed to maximize, inter-class variance proposed, while, the Kapur maximizes entropy to verify the homogeneity of the classes.

According to these functions, a large number of methods are proposed as image segmentation [10, 17, 18]. For example, Abd ElAziz et al. [19] proposed a modified version of SSO by using the FA for image segmentation. Also, the PSO and its modified versions are applied for image segmentation to locate the multi-level thresholding [20, 21].

Moreover, there are many other MH techniques have been applied for segmentation including harmony search (HS) algorithm [22], artificial bee colony (ABC) [23], honey bee mating optimization (HBMO) [24], and cuckoo search (CS) [25]. However, most of the previous mentioned MH techniques are based on maximizing/minimizing the single objective function. But, the ML thresholding image segmentation can be represented as a multiobjective optimization problem since this achieves many criteria at the same time [26]. The Multiobjective Optimization (MOP) methods aim to find the solutions that balance between these criteria and this represents the major difference between MOP and single-objective methods that offer a single solution. The set of solutions in a MOP is called the optimal Pareto front (PF), if they are not dominated by any other solutions. These PF solutions are providing the experts with the more suitable solution that balance between the different conditions.

Based on these concepts of MOPs, there are some multiobjective image segmentation methods are presented. For example, the non-dominated sorting genetic algorithm (NSGA II) has been proposed which maximizing the two objectives namely, the two-dimensional (2D) entropy and Shannon's entropy [27]. Based on the Kapur and Otsu functions, another multiobjective image segmentation method is proposed which called multiobjective particle swarm optimization (MOPSO) [28]. Also, the multiobjective evolutionary algorithm based on decomposition (MOEAD) based on cross-entropy and Renyi entropy [29]. However, most of these MOP image segmentation methods still required to be improved. Also, based on the No Free Lunch (NFL) theorem [30] which assumes that there doesn't exist one optimization method can solve all problems by the same quality. Therefore, this motivated us to provide an alternative MOP image segmentation method.

This chapter presents an alternative MOP image segmentation approach. This proposed method depends on the multiobjective grey wolf optimization (MOGWO) algorithm [31] that presented as an extension of the GWO algorithm [32]. The GWO simulates the behavior of the grey wolves to catch their preys in natural. These wolves are spread in three groups each of them has its task during the searching process about the prey. Based on these behaviors the traditional GWO algorithm has been applied to different applications including image segmentation, feature selection, cloud computing, and other applications. Moreover, the MOGWO algorithm is used to solve many real-world applications.

With all these advantages of using the MOGWO, it still not applied to image segmentation. So, this motivated us to used it to find the optimal threshold values by maximizing the two objective functions namely, the Kapur and Otsu. The proposed MOP image segmentation method starts by generating a set of solutions each of them represents the threshold levels required to optimize the objective function. Then the objective value for each solution is computed, and the current population is added to the archive that contains the non-dominated solutions. From this combination, the solutions in the archive are updated by determining the new non-dominated solutions and remove the other solutions from it. The next step is to select the best three solutions from the archive using the leader selection method and using these three solutions to update the other solutions in the population. Thereafter, computing the objective function for each updated solution and update the non-dominated solutions in the archive. Also, the size of the archive is checked if it becomes full or not. These previous steps are performed until reached to the existing conditions. The best threshold values can be obtained by selecting any solutions from the archive (exactly from the first front) and then compute the quality of this solution using the performance measures.

The rest this chapter is structured as follows: Sect. 13.2 explains the background of the problem dentition of multi-level thresholding, Kapur, Otsu functions, and the MOP. Section 13.3 introduces the proposed approach based on MOGWO. The experiments are introduced in Sect. 13.4. Finally, the conclusions and future works are explained in Sect. 13.5.

13.2 Problem Definition

Since the aim of this book is that each chapter can be analyzed by the reader separately the definition of the segmentation problem is also included in this chapter.

The multilevel thresholding is the process of finding the threshold levels (t) that split the image I into a set of K classes. This process can be formulated as follows:

$$C_0 = \{I_{ij}|0 \le I_{ij} < t_1 - 1\}$$
$$C_1 = \{I_{ij}|t_1 \le I_{ij} < t_2 - 1\}$$
$$\ldots$$
$$C_K = \{I_{ij}|t_K \le I_{ij} < L-\} \tag{13.1}$$

where $C_k(k = 1, 2, \ldots, K)$ represents the kth class which generated by using the threshold level t_k. I_{ij} is the pixel value of the image at position (i, j) and L is the total number of gray levels. However, to determine the threshold levels t_k they must be

maximizing (in some cases minimizing) suitable objective function (*Fit*) that defined
in the following equation:

$$t_1^*, t_2^*, \ldots, t_K^* = \arg \max_{1 \le k \le K} Fit(t_k) \tag{13.2}$$

The objective functions used in this study are the Otsu's and Kapur's entropy
which defined in the following.

13.2.1   Kapur's Entropy

The Kapur's entropy is considered one of the most popular image segmentation
approach which calculates the sum of entropies for the probability distribution of an
image histogram. Based on this definition the mathematical formula of the Kapur
functions is defined as [16]:

$$Fit_{Kapur} = \sum_{i=0}^{K} H_i, H_i = \sum_{j=t_i}^{t_{i+1}-1} \frac{\psi_j}{\gamma_j} \log\left(\frac{\psi_j}{\gamma_j}\right) \tag{13.3}$$

where the probability of the *j*th gray level is given by ψ_j.

13.2.2   Otsu's Function

In [15], another popular image segmentation method has been presented which called
Otsu's method. It is defined as the process of finding the threshold values that maximizing the following function:

$$Fit_{Otsu} = \sum_{i=0}^{K} \gamma_i(\mu_i - \mu_1)^2, \mu_i = \sum_{j=t_i}^{t_{i+1}-1} i\left(\frac{\psi_j}{\gamma_j}\right), \psi_i = \frac{h_i}{N} \tag{13.4}$$

$$\gamma_i = \sum_{j=t_i}^{t_{i+1}-1} \psi_j,$$

where h_i represents the frequency, μ_1 is the mean intensity of the image.

13.2.3 Multiobjective Optimization

The multiobjective problems (MOPs) have been applied to different applications such as clustering, classification, and several other pattern recognition applications [33, 34]. The main major of the MOPs is to provide the experts with a set of solutions which balance between different objectives and reduce the conflict between them.

In general, the mathematical formulation of the MOPs are presented by considering M objectives and the aim is to maximizing them as given in the following equation:

$$MaxFit(x) = \left[fit_1(x), fit_2(x), \ldots, fit_M(x)\right] \tag{13.5}$$

subject to:

$$g_i(x) = 0, i = 1, 2, \ldots, N_g, h_i(x) \geq 0, i = 1, 2, \ldots, N_h \tag{13.6}$$

where $fit_i(x)$ is the ith objective value of the solution x, whereas $g_i(x)$ and $h_i(x)$ are the equality and inequality constraint functions.

The process of determining the better solution is not easy like in the single objective function that compares between the solutions using the traditional operator such as relation operators. In MOP, the comparison is performed based on the concept of dominates, where the solution x is called dominates (better than) solution y when the following criteria is satisfied:

$$\forall i : fit_i(x) \geq fit_i(y), \wedge \exists j : fit_j(x) > fit_j(y), i, j \in \{1, 2, \ldots, M\} \tag{13.7}$$

Based on the criteria in Eq. (13.7), the set of solutions that not dominated by any other solution is called non-dominated (ND) solutions [or Pareto optimal (PO)] which represents the solutions to the MOPs. This set all ND is called Pareto set (PS), and it is defined as:

$$PS = \{x|x \text{ is } PO\} \tag{13.8}$$

The solutions of the Pareto set in objective space is called Pareto front (PF), and it is defined as:

$$PF = \{fit(x)|x \text{ is } PS \wedge fit(x) \in R^M\} \tag{13.9}$$

13.3 Multiobjective Grey Wolf Optimizer

In this section, the basic Grey Wolf Optimization (GWO) is introduced then the proposed multiobjective GWO algorithm for image segmentation.

13.3.1 Standard Grey Wolf Optimizer

The GWO has been produced by Mirjalili et al. [32] and takes its inspiration from the behavior of the living and the hunting of wolves. The main inspiration of the GWO algorithm is the social leadership and the technique of hunting of the grey wolves.

The social hierarchy of wolves can be mathematically modeled as following [32], the leader of the grey wolfs group can be noted as alpha (α), and it is considered as the fittest solution. Then the next levels in the group can be noted as beta (β) and delta (δ) wolves, which in turn can be considered as the second and third best solutions, respectively. The lower levels are omega (ω) wolves, where they are the rest of the candidate solutions.

The α, β, and δ can represent the optimization process in the GWO, and the ω wolves follow these three wolves in the search for the global optimum [32].

The mathematical model for the encircling behavior of the grey wolfs during the hunting process can be summarized as in the following equations [32]:

$$\vec{D} = \left| \vec{C}.\vec{X_p}(t) - \vec{X}(t) \right| \tag{13.10}$$

$$\vec{X}(t+1) = \vec{X_p}(t) - \vec{A}.\vec{D} \tag{13.11}$$

$$\vec{A} = 2\vec{a} \cdot \vec{r_1} - \vec{a}. \tag{13.12}$$

$$\vec{C} = 2 \cdot \vec{r_2} \tag{13.13}$$

where t represents the current iteration, $\vec{A}$ and $\vec{C}$ are coefficients, $\vec{X_p}$ represents the prey's position, and $\vec{X}$ represents a wolf's position. And the vector $\vec{a}$ has elements which are linearly decreasing from 2 to 0 through the iterations and $\vec{r_1}$, $\vec{r_2}$ are random vectors whose values are in the interval [0, 1].

The encircling technique of the GWO can, in turn, help it more in finding the optimization problems optimal solution.

This algorithm saves the higher levels solutions and obliges the lower levels solutions for updating the positions concerning them. The process of hunting can be summarized as following [32]:

$$\vec{D}_\alpha = \left| \vec{C}_1.\vec{X}_\alpha - \vec{X} \right| \tag{13.14}$$

$$\vec{D}_\beta = \left| \vec{C}_1.\vec{X}_\beta - \vec{X} \right| \tag{13.15}$$

$$\vec{D}_\delta = \left| \vec{C}_1.\vec{X}_\delta - \vec{X} \right| \tag{13.16}$$

$$\vec{X}_1 = \vec{X}_\alpha - \vec{A}_1 \cdot \left(\vec{D}_\alpha\right) \tag{13.17}$$

$$\vec{X}_2 = \vec{X}_\beta - \vec{A}_2 \cdot \left(\vec{D}_\beta\right) \tag{13.18}$$

$$\vec{X}_3 = \vec{X}_\delta - \vec{A}_3 \cdot \left(\vec{D}_\delta\right) \tag{13.19}$$

$$\vec{X}(t+1) = \frac{\vec{X}_1 + \vec{X}_2 + \vec{X}_3}{3} \tag{13.20}$$

where $\vec{C}$ is a vector of the exploration that can generate random values through the interval $[0, 2]$, and the prey weights stochastically emphasize $C > 1$ or deemphasize $C < 1$ [32]. The equation of $\vec{D}$ helps the GWO algorithm to have a random behavior more through optimization, in addition to avoiding the local optima. C is not linearly decreased when A is increased. When $|A| < 1$, the GWO algorithm's exploitation begins. If the $\vec{A}$ random values be in the interval $[-1, 1]$, then the following position of the agent may be in a position between the current position and the prey's position, By the alpha, beta, and delta solutions, the search agents can converge to an estimated position of the prey. The first step in the GWO algorithm is the generation of an initial population. Alpha, beta, and delta can be considered as the three best solutions. The last equations $(\vec{D}_\alpha, \vec{D}_\beta, \vec{D}_\delta)$ show the triggering for each omega wolf. Also, the coefficients $\vec{a}$, and $\vec{A}$ are decreased over iterations.

When $|A| > 1$, the wolf diverges away from the prey, whereas, as $|A| < 1$, the wolf converges to the prey [32].

13.3.2   Multi Objective GWO for Image Segmentation

In general, the MOGWO components that have been firstly introduced in [31] look like more those of the MOPSO. The first component is an archive for storing the obtained non-dominated Pareto optimal solutions. Also it can save and retrieve these solutions.

The second one is a leader selection strategy which can help in choosing alpha, beta, and delta solutions as the hunting process leaders which can be performed from the archive [31]. There is an archive controller which is considered as the key module of the archive, and it can control the archive as entering a solution into the archive or in case of the full archive, with a maximum number of the archive members. There is a comparison between the non-dominated solutions and the archive residents through iterations, where there are four cases [31]:

1. The agent cannot enter the archive if the new member is controlled by one resident of the archive at least.
2. Omit the dominated solution(s) in the archive if the new agent is a dominant of one or more solutions in the archive, then the new solution can enter the archive.
3. If neither the new solution nor archive members dominate each other, the new solution can enter the archive.
4. If the archive is full, the technique of the grid will be performed to make the objective space segmentation be arranged, also reach the segment which is the most crowded one to omit one of its solutions. After that, the new solution can enter the segment which is considered as the least crowded for enhancing the final approximated Pareto optimal front diversity. The omitting of a solution probability increases the number of solutions in the segment in a proportional way.

If the archive was full, omitting the solutions is performed by firstly selecting the segments that are considered the most crowded ones, then one solution is removed randomly, and a solution is omitted from one of them randomly for a space for the new one.

In the case of entering a solution away from the hypercubes, extend all the segments for covering the new solutions.

Another component is the technique of selecting the leader. Alpha, beta, and delta wolves can be three wolfs from the best solutions. And the leaders can help and guide the others.

The mathematical model of the MOGWO for image segmentation can be formulated by considering the tested image I is given as input, and its histogram. The next step is to give for each wolf from N wolves an initial position using the following equation:

$$X_i = I_{\min} + \sigma \times (I_{\max} - I_{\min})$$ (13.21)

where $I_{\min}$ and $I_{\max}$ represents the lower and upper boundaries for the search space. $\sigma \in [0, 1]$ represents a random number.

Then the objective function for each wolf is computed as using the following equation [31]:

$$F = \left[F_{Kapur} F_{Otsi} \right]^T$$ (13.22)

where the F_{Kapur} and F_{Otsu} are the Kapur and Otsu functions that defined in Eqs. (13.3) and (13.4), respectively.

Thereafter, determine the non-dominated solutions (NDS) and update the archive (AR). In addition, the leader selection component (LSC) is used to selects the least crowed segments of the search domain and provide one from the NDS as α, β, and δ solutions. This selection is performed using the following equation:

$$P_i = \frac{const}{Nseg_i} \qquad (13.23)$$

where *const* represents the constant number, and $Nseg_i$ represents the number of Pareto optimal solutions located in the ith segment.

Then the next step is to update the position of each wolf by using Equations (13.10)–(13.20). Then update the parameters C, A and a. The objective function is computed again for each solution and update the NDS of the archive AR. In addition, the size of the AR is checked before add any new NDS for it (as discussed previous). The previous steps are executed again until reached to the terminal conditions.

13.4   Experimental Results

The experiments are performed using a set of eight benchmark images which have different characteristics. Also, the results of the proposed MO method are compared with the other three algorithms namely, NSGA-II, MOPSO, and MOEAD. These algorithms are tested at threshold levels 2,3,5 (represents low levels), and 15(represents high level).

The structure of this section outlines as: The description of the eight images dataset is given in Sect. 13.4.1. In Sect. 13.4.2, performance metrics that used to assess the quality of the solution of each algorithm are introduced. The parameters of each method and the common parameters are explained in Sect. 13.4.3. Finally, the comparison results between proposed method and the other three methods are given in Sect. 13.4.4.

13.4.1   Dataset Description

The experiments performed on a set of popular natural images which used in many papers. This set of images contains eight grayscale images, which have the same size 512×512 except the third and fourth images which have size 256×256.

To illustrate the properties of each tested image the histogram of each one is depicted in Fig. 13.1. From this figure, it can be observed that some of these images contain more information than others like IM3. Whereas IM4 and IM7 are high- and low-key images, respectively.

13.4.1.1   Parameters Settings

In this study, three different multiobjective image segmentation method namely MOPSO [35], NSGA-II [36], and MOEAD [37] are used in comparison with the

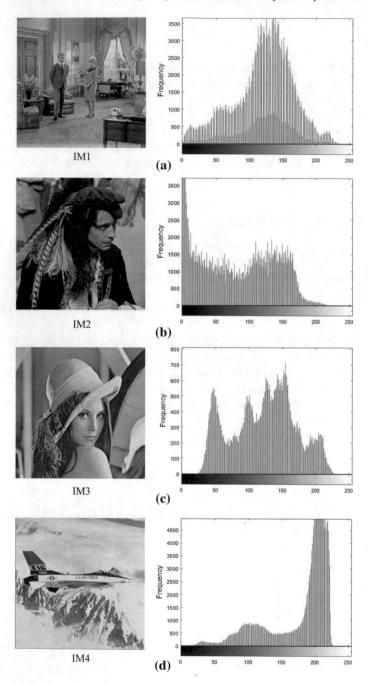

Fig. 13.1 The tested images and their histograms

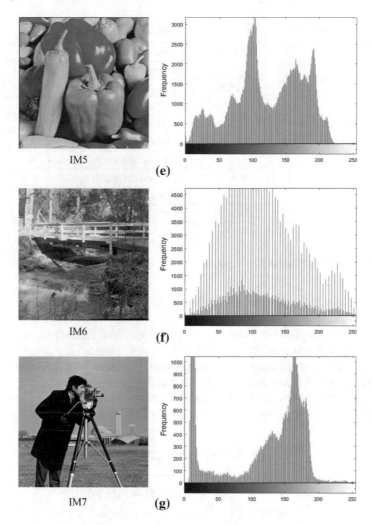

Fig. 13.1 (continued)

proposed method. The comparison is performed at low level threshold ($K = 2, 3, 5$) and at high level threshold $K = 15$.

In general, the common parameters between these methods are set as the following: the size of population is set 20 and the maximum number of iteration is to 500. In addition, each algorithm is repeated 20 times for statistical analysis. Meanwhile, the other parameters are set as in Table 13.1.

Table 13.1 The parameters settings for all algorithms

Method	Parameter	Value
MOEAD	Mutation scale factor	0.5
	Crossover constant	0.3
MOPSO	Cognitive coefficients	1.429
	Maximum velocity	1.0
	Minimum velocity	−1.0
	Minimum Inertia weight	0.4
	Maximum Inertia weight	0.9
NSGA-II	Mutation scale factor	0.5
	Crossover constant	0.3
MOGWO	a	1.5

13.4.2   Performance Measures

There are two sets of measures are used to compute the performance of each multi-objective image segmentation method. The first set aims to compute the quality of the segmented image and this including the PSNR, SSIM, fitness function, and the CPU time(s). Whereas, the second set aims to compute the quality of the algorithm to find the approximate Pareto front solutions and it includes the hypervolume (HV), and Spacing. These measures are defined in the following.

Peak-to-signal ratio (PSNR): It is used to compute the similarity between the original image I_{Gr} and its segmented image I_{th} as defined in the following equation:

$$PSNR = 20 \log_{10}\left(\frac{255}{RMSE}\right), \text{ (dB)}$$

$$RMSE = \sqrt{\frac{\sum_{i=1}^{ro}\sum_{j=1}^{co}(I_{Gr}(i,j) - I_{th}(i,j))}{ro \times co}} \quad (13.24)$$

where $RMSE$ represents the root mean square error. The ro, and co represents the total number of rows and columns of I_{Gr}, respectively. The algorithm that has higher PSNR value is the better algorithm.

Structure Similarity Index ($SSIM$): It is used to calculate the structures of I_{Gr} and I_{th} and it is defined as the following:

$$SSIM(I_{Gr}, I_{th}) = \frac{\left(2\mu_{I_{Gr}}\mu_{I_{th}} + C1\right)\left(2\sigma_{I_{Gr}I_{th}} + C2\right)}{\left(\mu_{I_{Gr}}^2 + \mu_{I_{th}}^2 + C1\right)\left(\sigma_{I_{Gr}}^2 + \sigma_{I_{th}}^2 + C2\right)}$$

$$\sigma_{I_oI_{Gr}} = \frac{1}{N-1}\sum_{i=1}^{N}\left(I_{Gr_i} + \mu_{I_{Gr}}\right)\left(I_{th_i} + \mu_{I_{th}}\right) \quad (13.25)$$

where $\mu_{I_{th}}(\sigma_{I_{th}})$ and $\mu_{I_{Gr}}(\sigma_{I_{Gr}})$ represent the mean (standard deviation) value of I_{th} and I_{Gr}, respectively. $C1$ and $C2$ are constants which both of them has value 0.065. The algorithm that has higher SSIM value is the better algorithm.

Hypervolume (HV) measure [38]: It is computing the area's size that dominated by other solutions in PF, then used it to measure the diversity and the closeness approximate PF. The HV is defined as the follows: $HV = volume(\cup_{i=1}^{|PF|}\{|PF|\}PF_i$. The Pareto set X is better than another solution Y if $HV(X)$ is greater than $HV(Y)$.

Spacing (SP) [38]: It measures the distribution of the non-dominated set:

$$SP = \sqrt{\frac{1}{\bar{N}}\sum_{i=1}^{\bar{N}}(d_i - \bar{d})^2} \tag{13.26}$$

where $\bar{N}$ represents is the number of solutions in PF. the d_i represents the Euclidean distance between x_i and its nearest solutions in the true Pareto front.

The algorithm that has the smallest SP value is the best.

13.4.3   Result and Discussion

The comparison results between the proposed method and the MOP methods are given in Figures and Tables that show the accuracy regarding SSIM, PSNR, HV, and SP, respectively.

According to the results of SSIM as in Table 13.2 and Fig. 13.2, it can be concluded that the MOGWO algorithm has the best results nearly overall the tested threshold levels (i.e., $K = 2, 3, 5$, and 15). In addition, it can see that the MOEAD provides better results than the NSGA-II and MOPSO at the low-level threshold levels $K = 2$, and 3. Meanwhile, at the other threshold levels t,he NSGA-II is better than MOEAD. Also, the MOPSO gives the worst results at all tested threshold levels except at the threshold level $K = 3$, which has SSIM value better than the NSGA-II.

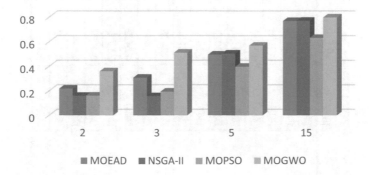

Fig. 13.2   The average of SSIM along each threshold

Table 13.2 Comparison results of the SSIM values for each algorithm

K	Img	MOEAD	NSGA-II	MOPSO	MOGWO	K	MOEAD	NSGA-II	MOPSO	MOGWO
2	IM1	0.1758	0.1596	0.1035	0.3381	5	0.5278	0.5340	0.4953	0.5955
	IM2	0.2129	0.1418	0.2312	0.3123		0.5579	0.5752	0.3738	0.5720
	IM3	0.2195	0.2739	0.1110	0.3054		0.6022	0.3976	0.2283	0.5052
	IM4	0.1231	0.1801	0.1076	0.2711		0.3770	0.3552	0.2754	0.5010
	IM5	0.3446	0.1221	0.2225	0.6298		0.4264	0.5698	0.5082	0.5987
	IM6	0.3017	0.1580	0.1119	0.2968		0.5030	0.5865	0.4005	0.6006
	IM7	0.1611	0.1120	0.2109	0.4490		0.5186	0.5016	0.3683	0.5594
	IM8	0.2117	0.1346	0.1929	0.2794		0.4524	0.5149	0.5299	0.6173
3	IM1	0.2675	0.1878	0.1560	0.6435	15	0.8208	0.7987	0.7176	0.8103
	IM2	0.2659	0.1861	0.2973	0.5623		0.7892	0.8109	0.5534	0.7837
	IM3	0.2423	0.0446	0.1310	0.5832		0.8663	0.7237	0.5431	0.8325
	IM4	0.1513	0.2454	0.1169	0.3528		0.6194	0.7940	0.5826	0.7443
	IM5	0.4229	0.0965	0.2603	0.5536		0.7564	0.8077	0.7749	0.8240
	IM6	0.4444	0.1813	0.2012	0.5560		0.7717	0.7508	0.6408	0.7872
	IM7	0.2569	0.1636	0.1595	0.4511		0.7494	0.7595	0.5851	0.8090
	IM8	0.4103	0.1510	0.2285	0.3844		0.7802	0.7195	0.6577	0.7885

Fig. 13.3 Average of the
PSNR values

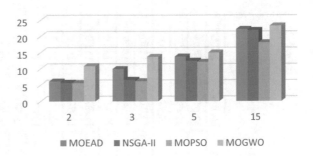

Regarding the results of the PSNR values that given in Table 13.3 and Fig. 13.3 it can be concluded that MOEAD algorithm has better results than the NSGA-II and MOPOS at all the threshold levels. However, the difference between them is decreased with the increase of the threshold level. Also, the proposed method has the first rank, in terms of PSNR value at all levels, and this indicates the high similarity between the segmented image by the proposed and the original image.

The results of evaluate the performance of the proposed method against the others in terms of the MOP measures are given in Tables 13.4 and 13.5. From Table 13.4, it can be observed that the algorithms based on the HV values have nearly the same performance. However, at threshold level 2, the MOGWO allocates the first rank followed by the NSGA-II, and MOPSO in second and fourth rank. Meanwhile, at the second threshold level, the NSGA-II allocates the first rank followed by the MOEAD and MOGWO in the second and third rank. The same conclusion can be reached at threshold level except the MOGWO is better than MOEAD. At the threshold level 15, the proposed MOGWO method has better results than other three algorithms, as well as, the MOEAD has better than the NSGA-II and MOPSO. In addition, the MOPSO has the worst results among all algorithms at the test threshold levels (see Fig. 13.4).

From Table 13.5 and Fig. 13.5, it can be observed the NSGA-II has better results than other three algorithms at threshold level 2. Whereas, at the threshold levels 3 and 5, MOGWO and the NSGA-II have the best SP values among the others algorithms. Moreover, the proposed MOGWo at the threshold level 15 allocates the first rank, then the MOPSO which allocates the second rank followed by the MOEAD and the NSGA-II.

13.4.4 Statistical Analysis

In this section, the statistical test called Wilcoxon's rank sum is used to study if there exist a significant difference between the proposed MOGWO and other algorithms or not. In general, Wilcoxon's rank sum test is a nonparametric test that doesn't depend on parameters like ANOVA test, therefore, it is more robust than ANOVA. There are two statistical hypotheses; the first hypothesis is called the null hypothesis which

Table 13.3 Comparison of the *PSNR* values between the algorithms

K	Img	MOEAD	NSGA-II	MOPSO	MOGWO	K	MOEAD	NSGA-II	MOPSO	MOGWO
2	IM1	5.75	5.67	5.65	10.06	5	13.47	10.61	12.66	14.14
	IM2	5.78	6.21	5.75	9.67		15.15	13.66	11.45	14.26
	IM3	6.34	5.51	5.47	10.29		15.30	12.54	9.55	13.60
	IM4	9.74	8.43	8.38	12.43		15.08	13.93	12.71	15.91
	IM5	3.13	2.89	2.96	11.91		9.28	12.93	13.54	13.77
	IM6	6.78	5.68	5.62	9.60		13.97	10.84	11.87	15.99
	IM7	5.93	6.01	5.93	13.54		14.89	13.61	11.60	15.97
	IM8	5.54	5.20	5.11	9.15		13.01	11.56	13.62	16.15
3	IM1	8.68	6.40	7.75	17.11	15	22.84	23.10	20.12	24.22
	IM2	9.15	7.38	5.75	14.21		22.35	23.51	16.07	22.04
	IM3	9.38	6.24	5.47	15.27		24.16	19.01	15.31	22.42
	IM4	10.25	9.41	8.38	12.72		21.87	25.46	20.32	23.22
	IM5	7.78	3.53	2.81	11.35		19.74	21.08	19.91	22.67
	IM6	11.89	6.50	5.62	13.71		21.53	21.27	18.29	23.76
	IM7	9.56	7.36	8.77	13.59		21.68	21.39	16.33	24.18
	IM8	12.80	6.23	5.11	11.57		23.67	20.38	18.31	23.61

Table 13.4 The results of HV for each algorithm among the tested images

K	Img	MOMVO	MOEADR	MOEAD	MOPSO	K	Img	MOMVO	MOEADR	MOEAD	MOPSO
2	IM1	1.01E+6	5.91E+5	2.38E+6	1.01E+6	2	IM2	2.65E+5	5.50E+5	1.59E+6	1.05E+6
3		1.59E+6	1.85E+6	2.62E+6	3.16E+6	3		3.10E+6	3.14E+6	3.75E+6	1.27E+6
5		2.23E+6	2.02E+6	2.75E+6	3.85E+6	5		3.89E+6	3.17E+6	4.02E+6	2.71E+6
15		2.38E+6	3.13E+6	2.99E+6	4.22E+6	15		4.35E+6	3.40E+6	4.05E+6	4.01E+6
2	IM3	3.38E+6	4.04E+5	6.11E+5	3.82E+5	2	IM4	1.37E+6	1.03E+6	1.15E+6	1.57E+6
3		3.38E+6	6.37E+5	3.64E+6	3.89E+5	3		3.57E+6	1.36E+6	1.60E+6	3.42E+6
5		4.26E+6	2.86E+6	4.09E+6	1.18E+6	5		3.59E+6	1.67E+6	2.22E+6	3.71E+6
15		4.80E+6	4.43E+6	4.89E+6	3.45E+6	15		4.48E+6	4.63E+6	3.58E+6	4.14E+6
2	IM5	4.61E+5	1.49E+6	2.04E+5	1.57E+6	2	IM6	4.74E+5	1.80E+6	1.65E+6	1.04E+6
3		1.28E+6	4.22E+6	1.95E+6	1.93E+6	3		2.52E+6	3.61E+6	2.19E+6	1.35E+6
5		1.32E+6	4.81E+6	2.99E+5	2.18E+6	5		3.77E+6	3.92E+6	2.99E+6	3.22E+6
15		4.61E+6	4.83E+6	4.76E+6	3.59E+6	15		3.79E+6	4.06E+6	4.14E+6	4.47E+6
2	IM7	2.79E+5	2.16E+5	3.40E+5	8.06E+5	2	IM8	4.59E+5	2.01E+6	6.32E+5	9.59E+5
3		1.55E+6	1.53E+6	2.65E+6	2.17E+6	3		1.31E+6	2.39E+6	1.38E+6	1.50E+6
5		2.04E+6	1.63E+6	2.73E+6	2.23E+6	5		2.26E+6	2.96E+6	2.54E+6	2.27E+6
15		3.06E+6	4.70E+6	3.28E+6	3.07E+6	15		3.07E+6	3.38E+6	4.01E+6	2.35E+6

Table 13.5 The results of the S for each algorithm at the tested images

K	Img	MOGWO	MOEAD	NSGA-II	MOPSO	Img	MOGWO	MOEAD	NSGA-II	MOPSO
2	IM1	0.0048	0.0060	0.0037	0.0052	IM2	0.0058	0.0049	0.0050	0.0043
3		0.0031	0.0057	0.0030	0.0046		0.0056	0.0046	0.0040	0.0029
5		0.0014	0.0044	0.0017	0.0040		0.0055	0.0046	0.0039	0.0013
15		0.0013	0.0036	0.0010	0.0029		0.0033	0.0031	0.0031	0.0011
2	IM3	0.0056	0.0037	0.0050	0.0055	IM4	0.0046	0.0046	0.0029	0.0053
3		0.0035	0.0029	0.0032	0.0049		0.0025	0.0040	0.0026	0.0047
5		0.0029	0.0017	0.0025	0.0045		0.0020	0.0017	0.0026	0.0023
15		0.0013	0.0011	0.0023	0.0027		0.0012	0.0011	0.0015	0.0017
2	IM5	0.0056	0.0050	0.0049	0.0052	IM6	0.0054	0.0047	0.0042	0.0051
3		0.0043	0.0050	0.0048	0.0050		0.0033	0.0043	0.0041	0.0048
5		0.0026	0.0016	0.0037	0.0013		0.0032	0.0037	0.0039	0.0048
15		0.0016	0.0012	0.0025	0.0012		0.0015	0.0020	0.0036	0.0021
2	IM7	0.0039	0.0059	0.0054	0.0054	IM8	0.0051	0.0046	0.0057	0.0057
3		0.0039	0.0046	0.0031	0.0031		0.0040	0.0042	0.0051	0.0045
5		0.0024	0.0025	0.0024	0.0015		0.0020	0.0034	0.0046	0.0025
15		0.0011	0.0016	0.0019	0.0013		0.0016	0.0021	0.0027	0.0021

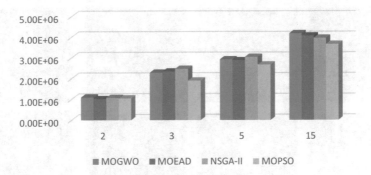

Fig. 13.4 The average of HV value for each algorithm

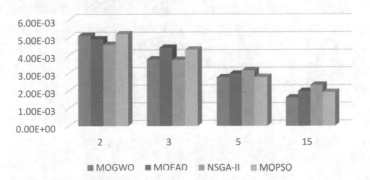

Fig. 13.5 The average of SP value overall the images at each threshold level

Table 13.6 The Wilcoxon's rank results

Algorithm	Value	PSNR	SSIM	HV	SP
MOEAD	P	0.2400	0.2455	0.9411	0.4681
	H	0	0	0	0
NSGA-II	P	0.0639	0.0574	0.8773	0.6771
	H	0	0	0	0
MOPSO	P	0.0233	0.0213	0.3612	0.6575
	H	1	1	0	0

assumes that there is no significant difference between the proposed method and the other methods. Meanwhile, another hypothesis is called the alternative hypothesis that assumes that there is a significant difference. The process of accepting the null hypothesis (which mean reject the alternative) depends on the statistical value (P), if this value is greater than 0.05 then accept this hypothesis (i.e., H = 0), otherwise, reject it (i.e., H = 1).

Table 13.6 shows the results of Wilcoxon's rank sum test; from this table it can be observed that there is no significant difference between the proposed method and the other in terms of the MOP indicators (i.e., HV and SP). As well as, based on the PSNR,

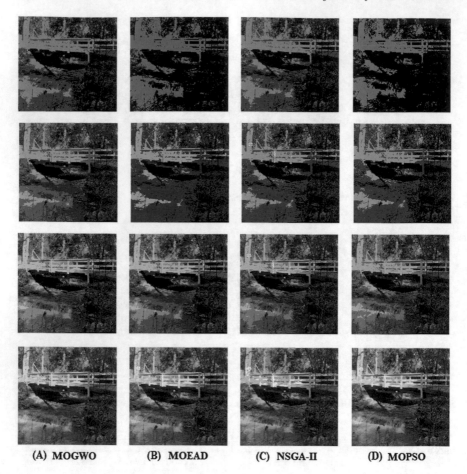

(A) MOGWO   (B) MOEAD   (C) NSGA-II   (D) MOPSO

Fig. 13.6 The segmented image using each algorithm at thresholds (K = 2 in first row and K = 15 last row)

and SSIM, the null hypothesis is accepted when the proposed MOGWO method is compared with the MOEAD and NSGA-II. However, the alternative hypothesis is accepted when MOGWO is compared with the MOPSO based on PSNR and SSIM (Fig. 13.6).

From all the previous results it can be concluded that the proposed method has good quality compared to the three algorithm. However, there is some limitations of the MOGWO method, such as there are some parameters that need to determine at the beginning of executed the MOGWO.

13.5   Conclusions

This chapter introduced an alternative multiobjective optimization image segmentation method. This proposed method depends on using the multiobjective grey wolf optimization (MOGWO) that considered as an extension of the traditional GWO algorithm. The GWO algorithm simulates the behavior of grey wolves to catch the prey. These behaviors are formulated as a mathematical optimization model which applied in different applications. However, this model is applied only to the single objective optimization problem. Therefore, it is an extension that called MOGWO is proposed to deal with multiobjective optimization problems (MOPs). This MOGWO has been applied to different applications but not applied to image segmentation yet. In this chapter, the multiobjective image segmentation based on Kapur and Otsu function is proposed by using MOGWO as an optimization algorithm. In order to evaluate the performance of the MOGWO, a set of experiments are performed using eight images. In addition, the accuracy of MOGWO is compared with other multiobjective algorithms namely, NSGA-II, MOEAD, and MOPSO at different four threshold levels. From this experiment, the proposed MOGWO provide results better than the other three algorithms at the most of threshold levels regarding the quality of segmented image and MOP indicators.

Based on the previous results of MOGWO, it can be applied in the future for color image segmentation, medical image, and several other image processing applications.

References

1. Maitra M, Chatterjee A (2008) A novel technique for multilevel optimal magnetic resonance brain image thresholding using bacterial foraging. Meas J Int Meas Confed 41:1124–1134. https://doi.org/10.1016/j.measurement.2008.03.002
2. Ortiz A, Górriz JM, Ramírez J, Salas-González D (2013) Improving MRI segmentation with probabilistic GHSOM and multiobjective optimization. Neurocomputing 114:118–131. https://doi.org/10.1016/j.neucom.2012.08.047
3. Liu C, Ng MK, Zeng T (2018) Weighted variational model for selective image segmentation with application to medical images. Pattern Recognit 76:367–379. https://doi.org/10.1016/j.patcog.2017.11.019
4. Hamuda E, Glavin M, Jones E (2016) A survey of image processing techniques for plant extraction and segmentation in the field. Comput Electron Agric 125:184–199. https://doi.org/10.1016/j.compag.2016.04.024
5. Demant C, Streicher-Abel E, Garnica C. Industrial Image Processing, Visual Quality Control in Manufacturing, Springer-Verlag Berlin Heidelberg 2013, ISBN 978-3-642-33904-2
6. Pathak S, Sejwar V (2017) A review on image segmentation using different optimization techniques. International Journal of computer sciences and Engineering 217–221
7. Li H, He H, Wen Y (2015) Dynamic particle swarm optimization and K-means clustering algorithm for image segmentation. Opt—Int J Light Electron Opt 126:4817–4822. https://doi.org/10.1016/j.ijleo.2015.09.127
8. Detection ME (2005) A Contour based Image segmentation algorithm using morphological edge detection. In: 2005 IEEE International Conference on systems, man and cybernetics, pp 2962–2967

9. Mittal H H, Saraswat M (2018) An optimum multi-level image thresholding segmentation using non-local means 2D histogram and exponential Kbest gravitational search algorithm. Eng Appl Artif Intell 71:226–235. https://doi.org/10.1016/j.engappai.2018.03.001

10. El Aziz MA, Ewees AA, Hassanien AE (2017) Whale optimization algorithm and moth-flame optimization for multilevel thresholding image segmentation. Expert Syst Appl 83:242–256. https://doi.org/10.1016/j.eswa.2017.04.023

11. Kennedy J, Eberhart RC (1995) Particle swarm optimization, vol 4. In: 1995 Proceedings of IEEE international conference on neural networks, pp 1942–1948. https://doi.org/10.1109/icnn.1995.488968

12. Mirjalili S, Lewis A (2016) The whale optimization algorithm. Adv Eng Softw 95:51–67. https://doi.org/10.1016/j.advengsoft.2016.01.008

13. Yang S, Cheng H, Wang F (2010) Genetic algorithms with immigrants and memory schemes for dynamic shortest path routing problems in mobile *Ad Hoc* networks. IEEE Trans Syst Man Cybern Part C (Appl Rev) 40:52–63

14. Ibrahim RA, Ewees AA, Oliva D, Elaziz M, Songfeng L (2018) Improved salp swarm algorithm based on particle swarm optimization for feature selection. J Ambient Intell Human Comput 1–15. https://doi.org/10.1007/s12652-018-1031-9

15. Otsu N (1979) A threshold selection method from gray-level histograms. IEEE Trans Syst Man Cybern 9:62–66. https://doi.org/10.1109/TSMC.1979.4310076

16. Kapur JN, Sahoo PK, Wong AK (1985) A new method for gray-level picture thresholding using the entropy of the histogram. Comput Vis Graph Image Process 29:273–285

17. Bhandari AK, Kumar A, Chaudhary S, Singh GK (2016) A novel color image multilevel thresholding based segmentation using nature inspired optimization algorithms. Expert Syst Appl 63:112–133. https://doi.org/10.1016/j.eswa.2016.06.044

18. Manikandan S, Ramar K, Iruthayarajan MW, Srinivasagan K (2014) Multilevel thresholding for segmentation of medical brain images using real coded genetic algorithm. Measurement 47:558–568

19. Bhattacharyya S, Dutta P, De S, Klepac G (2016) Hybrid soft computing for image segmentation. Hybrid Soft Comput Image Segmentation 1–321. https://doi.org/10.1007/978-3-319-47223-2

20. Ghamisi P, Couceiro MS, Benediktsson JA, Ferreira NM (2012) An efficient method for segmentation of images based on fractional calculus and natural selection. Expert Syst Appl 39:12407–12417

21. Nakib A, Roman S, Oulhadj H, Siarry P (2007) Fast brain MRI segmentation based on two-dimensional survival exponential entropy and particle swarm optimization. In: 29th annual international conference of the IEEE in engineering in medicine and biology society. EMBS, pp 5563–5566

22. Oliva D, Cuevas E, Pajares G, et al (2013) Multilevel thresholding segmentation based on harmony search optimization. J Appl Math. https://doi.org/10.1155/2013/575414

23. Akay B (2013) A study on particle swarm optimization and artificial bee colony algorithms for multilevel thresholding. Appl Soft Comput J 13:3066–3091. https://doi.org/10.1016/j.asoc.2012.03.072

24. Horng MH, Liou RJ (2011) Multilevel minimum cross entropy threshold selection based on the firefly algorithm. Expert Syst Appl 38:14805–14811. https://doi.org/10.1016/j.eswa.2011.05.069

25. Pare S, Kumar A, Bajaj V, Singh GK (2016) A multilevel color image segmentation technique based on cuckoo search algorithm and energy curve. Appl Soft Comput J 47:76–102. https://doi.org/10.1016/j.asoc.2016.05.040

26. Ramezani F, Lu J, Taheri J, Hussain FK (2015) Evolutionary algorithm-based multi-objective task scheduling optimization model in cloud environments. World Wide Web 18:1–23. https://doi.org/10.1007/s11280-015-0335-3

27. Nakib A, Oulhadj H, Siarry P (2010) Image thresholding based on Pareto multiobjective optimization. Eng Appl Artif Intell 23:313–320. https://doi.org/10.1016/j.engappai.2009.09.002

28. Yin PY, Wu TH (2017) Multi-objective and multi-level image thresholding based on dominance and diversity criteria. Appl Soft Comput J 54:62–73. https://doi.org/10.1016/j.asoc.2017.01.019

29. Sarkar S, Das S, Chaudhuri SS (2017) Multi-level thresholding with a decomposition-based multi-objective evolutionary algorithm for segmenting natural and medical images. Appl Soft Comput J 50:142–157. https://doi.org/10.1016/j.asoc.2016.10.032

30. Wolpert DH, Macready WG (1997) Simple explanation of the no free lunch theorem of optimization. IEEE Trans Evol Comput 1:67–82. https://doi.org/10.1109/.2001.980896

31. Mirjalili S, Saremi S, Mirjalili SM, Coelho LS (2015) Multi-objective grey wolf optimizer: a novel algorithm for multi-criterion optimization. Expert Syst Appl 47:106–119. https://doi.org/10.1016/j.eswa.2015.10.039

32. Mirjalili S, Mirjalili SM, Lewis A (2014) Grey wolf optimizer. Adv Eng Softw 69:46–61. https://doi.org/10.1016/j.advengsoft.2013.12.007

33. Jiang Q, Wang L, Lin Y et al (2017) An efficient multi-objective artificial raindrop algorithm and its application to dynamic optimization problems in chemical processes. Appl Soft Comput J 58:354–377. https://doi.org/10.1016/j.asoc.2017.05.003

34. El Aziz MA, Ewees AA, Hassanien AE et al (2018) Multi-objective whale optimization algorithm for multilevel thresholding segmentation. Stud Comput Intell 730:23–39. https://doi.org/10.1007/978-3-319-63754-9_2

35. Zhang X, Tian Y, Cheng R, Jin Y (2105) An efficient approach to nondominated sorting for evolutionary multiobjective optimization. IEEE Trans Evol Comput 19:201–213

36. Chen X, Du W, Qian F (2014) Multi-objective differential evolution with ranking-based mutation operator and its application in chemical process optimization. Chemom Intell Lab Syst 136:85–96. https://doi.org/10.1016/j.chemolab.2014.05.007

37. Zhang Q, Li H (2007) MOEA/D: A multiobjective evolutionary algorithm based on decomposition. 11:712–731. https://doi.org/10.1109/tevc.2007.892759

38. Zitzler E, Thiele L (1999) Multiobjective evolutionary algorithms: a comparative case study and the strength Pareto approach. IEEE Trans Evol Comput 3:257–271

Chapter 14
Clustering Algorithms for Image Segmentation

14.1 Introduction

Clustering is a tool from machine learning that helps to separate the information from a dataset in different groups. It works classifying the elements of the dataset based on the distance that exists between them and a centroid. The output from a clustering algorithm is basically a statistical description of the cluster centroids with the number of components in each cluster. Figure 14.1 shows an easy example of clustering. From this figure, it is easy two see that the data is distributed in a two-dimensional space. Moreover, such data creates three different clusters that can be easily detected by humans. Then the problem is for computers because the algorithms are not able to detect the information like humans. For that reason, the cluster algorithms must be able to work with all the information using rules that permits to classify the information in different groups.

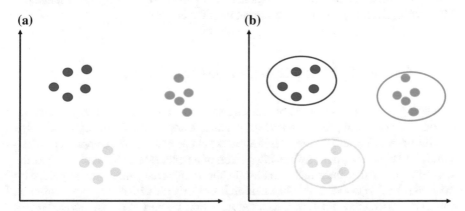

(a) **(b)**

Fig. 14.1 **a** Scattered data, **b** Clustered data

© Springer Nature Switzerland AG 2019

D. Oliva et al., *Metaheuristic Algorithms for Image Segmentation:*
Theory and Applications, Studies in Computational Intelligence 825,
https://doi.org/10.1007/978-3-030-12931-6_14

Since clustering aims to divide the data into groups with similar objects, each group has dissimilarities with other groups [1]. Clustering methods have some points that should be considered to be properly implemented. They are the scalability that permits to be used on different qualities and the ability to avoid the noise contained in the data (interpretability and usability) [2].

In image processing, it is an important method that helps to classify the pixels according to their intensity. The advantage of using clustering for image segmentation is that is not necessary to work with the histogram. It means that the distribution of pixels must be considered in a bounded space that could be bi-dimensional or tri-dimensional. One of the main problems in clustering is the initialization of the centroids that defines the classes [3]. For an accurate classification of the pixels, the centroids must be accurately selected. In standard versions, this process is performed by an exhaustive search. This process is computationally expensive and the complexity increases depending on the data and the number of classes.

The metaheuristic optimization algorithms have introduced to overcome the problems generated for searching the best configuration of centroids in clustering. There exist different implementations in the related literature for image segmentation with metaheuristics. They can be applied for both grayscale and color images; here the problem is to select the best optimization approach that provides better performance for the set of images that is processed. Moreover, in the state-of-the-art different clustering techniques have been introduced. The classics are the K-means and C-means. However, there is possible to find more sophisticated approaches that include the use of fuzzy sets to avoid the noise that affects the classification of the information.

This chapter aims to provide the theory that permits the reader the implementation of different methods for image segmentation using clustering. Here are explained different objective functions that could be easily adapted to be optimized by an optimization algorithm.

The rest of the sections are distributed as follows: Sect. 14.2 presents the basic concepts of the clustering methods that are commonly used for image segmentation. Meanwhile, Sect. 14.3, provides a summary of the entire chapter.

14.2   Clustering Methods for Image Segmentation

Clustering is part of the machine learning methods, and it is used to partitioning the data into different groups. As part of machine learning, clustering can be divided into two categories namely supervised approaches and unsupervised approaches. The supervised clustering requires the interaction of an expert to guide and verify if the separation of the data is correct. Meanwhile, the unsupervised methods are guided by internal rules that decide if the information is properly classified. The most common clustering approaches used for image segmentation are part of the supervised methods. Some examples include the hierarchical approaches such as relevance feedback techniques [4]. The clustering techniques which are included in this paper are Hierarchical clustering [5], Partitional clustering [6], K-means clustering [7] and Fuzzy clustering [8].

The first methods explained in this section are briefly introduced since they are part of classical approaches and the reader can explore more about them in the related literature. However, for the approaches that are extensively used their explanations are more extensive.

14.2.1 Hierarchical Clustering

The first technique treated in this chapters it the hierarchical clustering [9, 10]. It is part of information retrieval, and it is also extensively used for classification of complex datasets. The hierarchical clustering works considering a proximity matrix that is used to verify the similarity between each pair of data that are clustered. This methodology generates a tree of clusters that represent nested groups of patterns that also contains different similarity levels. These groups are created as an internal node of the tree. The differences between the clustering algorithms is in the rules that they use to merge two small clusters or for split large clusters. In general terms the hierarchical clustering approaches can be divided in agglomerative and divisive. Agglomerative algorithms request to merge clusters to be larger by starting with N single point clusters. This algorithm can be divided into three groups:

1. Single link algorithm
2. Complete link algorithm
3. Minimum variance algorithm.

In the first group, the single link algorithm works merging two clusters using the minimum distance that exist between the data samples of two different clusters. For the second group, the complete link algorithm also employs the maximum distance that is present in the data samples. The third group called the minimum variance algorithm, the methodology employed combines two clusters to minimize the cost function and creates a new cluster.

On the other hand, the divisive clustering starts using the same cluster in the entire dataset. It follows a reverse splitting of the dataset until the single point clusters are attained on leaf nodes. The methodology that it follows employs a reverse clustering approach that is completely different to the agglomerative algorithm.

To explain the standard process of hierarchical clustering here is introduced a simple example. Consider a set of N items to be clustered, and an $N*N$ distance (or similarity) matrix, then follow the next steps for cluster the data:

Step 1: Start by assigning each item to a cluster. It means that for N items, there will be N clusters and each of them contain just one item. Let the distances (similarities) between the clusters the same as the distances (similarities) between the items they contain.

Step 2: Find the closest (most similar) pair of clusters and merge them into a single cluster. Now the number of clusters is reduced in 1.

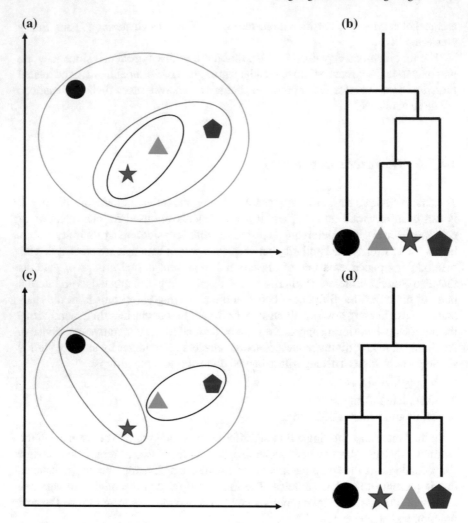

Fig. 14.2 **a** Hierarchical clustering at 1 iteration, **b** dendrogram at 1 iteration, **c** Hierarchical clustering at 2 iterations, **d** dendrogram at 2 iterations

Step 3: Compute distances (similarities) between the new cluster and each of the old clusters.

Step 4: Repeat steps 2 and 3 until all items are clustered into a single cluster of size N.

Step 3 can be done in different ways, which is what distinguishes single-link, complete-link, minimum variance algorithm, and average-linkage clustering. Figure 14.2 presents an example of hierarchical clustering.

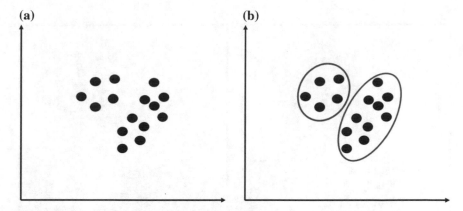

Fig. 14.3   a Scattered data, **b** partitional clustering

14.2.2   *Partitional Clustering*

The partitional clustering is another interesting method that works in an iterative way using optimization procedure. The goal of this technique is minimizing an objective function that is used to measure the quality of the clustering [6]. In this context, it is necessary to define the centroids of clusters. This task is commonly performed based on an optimality criterion that permits the proper minimization of the objective function. Figure 14.3 presents an example of the partitional clustering approach. In general, the partitional clustering algorithms can be divided into two categories:

1.   Partitioning relocation algorithm
2.   Density-based partitioning

14.2.3   *K-Means Clustering*

The K-means is considered one of the easiest and simplest method for supervised data classification [2]. This approach separates the data extracted from an image in clusters using an iterative process. Basically, it computes a mean intensity value for each class, and then each pixel is categorized in the class that possesses the closest mean. The clustering approaches that consider the optimization of an overall measure have extensively studied and are very attractive for different scientific communities like pattern recognition, machine learning, artificial intelligence, etc. The K-means is one of the most popular methods using for different implementations. It has been taken from the pattern recognition field. In the K-means it is computed a multidimensional centroid for each cluster. Such centroid is selected under the consideration that it must minimize the distance that exists between all the clusters [9]. The entire process for K-means is summarized in the following steps:

(a)   **(b)**

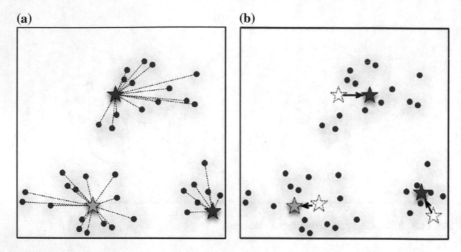

Fig. 14.4 **a** Cluster assignment, **b** cluster movement

Step 1: Generate the K initial clusters $z_1(s), z_2(s), \ldots, z_k(s)$

Step 2: For an iteration i-th, distribute the data samples x into the K clusters as follows:

$$x \in C_j(k) \text{ if } \|x - z_m(k)\| \leq \|x - z_n(k)\|$$

where $m = 1, 2, \ldots, k$ and $m \neq n$. $C_j(k)$ is a set of samples with a centroid defined as $z_n(k)$.

Step 3: Obtain the new centroids $z_n(k + 1), n = 1, 2, \ldots, K$. Consider that the sum of squared distances from all points in $C_j(k)$, then new clusters needs to be minimized. In this step it is used the mean of $C_j(k)$ as a metric that must be minimized. The new clusters are obtained as:

$$z_n(k + 1) = \frac{1}{N_n} \sum x, \, n = 1, 2, \ldots, K$$

where N_n is the number of samples contained in $C_j(k)$.

Step 4: Verify if all the data has been classified according to the centroids, if it occurs the algorithm converges, and the process ends, otherwise, go to Step 2.

The K-means algorithm is an iterative method that works until a stop criterion is achieved. This algorithm can be summarized in two tasks, namely: (1) cluster assignment and (2) cluster movement. In Fig. 14.4 is presented the process of K-means, where Fig. 14.4(a) represents the cluster assignment and Fig. 14.4(b) corresponds to cluster movement.

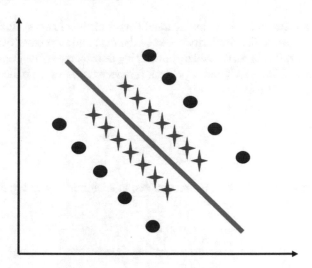

Fig. 14.5 An example of overlapped clusters, the red stars belong to two different clusters

14.2.4 Fuzzy Clustering

In clustering there exist some cases in which the data belong to more than one cluster, here the clusters are overlapped (see Fig. 14.5). In these situations, there is necessary to use specific kind of algorithms. In this context, the clustering algorithms can be divided into hard and fuzzy methods. Hard clustering employs a membership value of zero or one that is assigned to each pattern of the data. Meanwhile, the fuzzy clustering uses a value that exists between zero and one and by the use of a membership function it is assigned to the data. Some authors affirm that fuzzy methods are superior to hard approaches because they are able to represent a good relationship between data and clusters. During the clustering procedure, the fuzzy methods employ a heuristic function.

On the other hand, the fuzzy C-means (FCM) is an algorithm that iteratively updates the centers of the clusters and estimates the class membership function using the gradient descendent. This method is very popular, and it has several implementations in different fields. This method (developed by Dunn in 1973 and improved by Bezdek in 1981 [8, 11] is frequently used in pattern recognition. It is based on the minimization of the following objective function:

$$J_m = \sum_{i=1}^{N} \sum_{j=1}^{C} u_{ij}^m \|x_i - c_j\|^2, \ 1 \le m < \infty \tag{14.1}$$

From Eq. (14.1) m is a real number higher than 1. The degree of membership of x_i in the cluster j is represented by u_{ij}, x_i is the i-th element of the d-dimensional dataset. Meanwhile, c_j is the d-dimensional center of the cluster j, finally $\|\ \|$ is the norm used to measure the similarity between the data and the center of a cluster.

The fuzzy partition is performed by optimizing the objective function described in Eq. (14.1). Since FCM is an iterative process, it is necessary to update the membership function using Eq. (14.2).

$$u_{ij} = \frac{1}{\sum_{k=1}^{C} \left(\frac{\|x_i - c_j\|}{\|x_i - c_k\|} \right)^{2/m-1}} \qquad (14.2)$$

The centers of the clusters are also updated at each iteration using the following equation:

$$c_j = \frac{\sum_{i=1}^{N} u_{ij}^m \cdot x_i}{\sum_{i=1}^{N} u_{ij}^m} \qquad (14.3)$$

The FCM needs an stop criteria; it could be a predefined number of iterations, an error between the data classified, or a value of the objective function.

14.3 Summary

This chapter presents some of the most used clustering methods for image segmentation. These kind of techniques cannot be categorized into the thresholding approaches since they work directly over the pixels of the image. In other words, the pixels are considered as data that scattered over a search space.

The clustering methods are an iterative process that in most of the cases tries to optimize a function that measures the similarities between the data and the center of the cluster. Under such circumstances, it is easy to address the problems with metaheuristics. In most of the cases, the metaheuristic algorithms are employed to find the best centroids and the objective function could be a distance between the data and the centers.

Here is also important to mention that clustering can be used for both grayscale images and color images (e.g., RGB). For color images the idea is the same just the dimension of the problem change, instead to have two-dimensional data now the data has tree dimension one for each channel on the color space.

References

1. Rai P, Singh S (2010) A survey of clustering techniques. Int J Comput Appl 7:1–5. https://doi.org/10.5120/1326-1808
2. Saraswathi S, Allirani A (2013) Survey on image segmentation via clustering. In: International conference on information communication and embedded systems, ICICES 2013, pp 331–335
3. Anter AM, Hassenian AE, Oliva D (2019) An improved fast fuzzy c-means using crow search optimization algorithm for crop identification in agricultural. Expert Syst Appl 118:340–354. https://doi.org/10.1016/j.eswa.2018.10.009
4. Zhou XS, Huang TS (2003) Relevance feedback in image retrieval: a comprehensive review. Multimed Syst 8:536–544. https://doi.org/10.1007/s00530-002-0070-3
5. Johnson SC (1967) Hierarchical clustering schemes. Psychometrika 32:241. https://doi.org/10.1007/BF02289588
6. Jin X, Han J (2011) Partitional clustering. Encycl Mach Learn pp 32–33
7. Mac Queen J (1967) Some methods for classification and analysis of multivariate observations. In: Proceedings of the fifth Berkeley symposium on mathematical statistics and probability, pp 281–297
8. Dunn JC (1973) A fuzzy relative of the ISODATA process and its use in detecting compact well-separated clusters. J Cybern 3:32–57. https://doi.org/10.1080/01969727308546046
9. Thilagamani S, Shanthi N (2011) A survey on image segmentation through clustering algorithm. Int J Sci Res 1:14–17
10. Min H, Bo S, Jianqing X (2009) An optimized image retrieval method based on hierarchal clustering and genetic algorithm. In: Proceedings of 2009 international forum on information technology and applications, IFITA 2009, vol 1, pp 747–749. https://doi.org/10.1109/ifita.2009.429
11. Bezdek JC (1981) Pattern recognition with fuzzy objective function algorithms, 1st edn. Springer, Heidelberg

Chapter 15
Contextual Information in Image Thresholding

15.1 Introduction

As has been stated repeatedly on previous chapters, the threshold-based segmentation is a useful and simple tool in image processing, there being a wide diversity of techniques with continuous contributions. However, most of the proposed research focuses on how to select the threshold values on a given image taking in consideration the value of each pixel. Recently the concept of the energy curve was introduced as a thresholding method to consider the spatial information around pixels [1]. Consequently, the objective of this chapter focuses on this aspect by evaluating the results derived from the incorporation of the concept of energy curve in traditional and more recent metaheuristic algorithms.

Thresholding is an image segmentation technique that enjoys simplicity, robustness and precision [2, 3]. However, multi-level thresholding with classical methods creates problems since such implementations are often associated with a high computational cost to find threshold values. Image thresholding has been treated using metaheuristic algorithms to search for the best threshold values without greatly increasing the computational cost. Some recent contributions demonstrate the validity of this approach. The application of metaheuristic algorithms recently applied to the threshold problem are good examples of this [4–8].

Although histogram-based approaches are the most widely used, efforts have been made to incorporate spatial information into the process of threshold-based segmentation. A relevant technique that incorporated spatial information is the work of Ghosh [1], later adapted to the Cuckoo Search algorithm [9] and later expanded through the use of co-occurrence matrices for grey intensities [10]. As the incorporation of the energy curve into the threshold process is relatively new, there is still no comprehensive evaluation of the potential of this new methodology in the literature.

Under the previous considerations, this chapter analyses the incorporation of contextual information into the segmentation process by means of the energy curve. The chapter is organized as follows: in Sect. 15.1 the concept of energy curve is intro-

© Springer Nature Switzerland AG 2019
D. Oliva et al., *Metaheuristic Algorithms for Image Segmentation:
Theory and Applications*, Studies in Computational Intelligence 825,
https://doi.org/10.1007/978-3-030-12931-6_15

duced. Section 15.2 describes two algorithms of a contextual nature. Section 15.3 includes the analysis of the results obtained, which is completed by an assessment in Sect. 15.4.

15.2   Energy Curve

The energy curve is a means of representing the information contained in an image just like the histogram. The energy curve has interesting properties such as the incorporation of spatial or contextual information from the image and not just the intensity of each pixel, which is the key difference with respect to the histogram. Compared to histograms, the energy curves are visually smoother, conserving valleys and peaks. In this case, the image thresholding process consists of searching for the threshold values corresponding to the middle of valley regions in the energy curve. Each valley exists between two adjacent modes, each mode characterizes an object in the image. Thus, the objective is the precise selection of threshold values to obtain the best possible results in segmentation.

The image $I = \{l_{ij}, 1 \le i \le m, 1 \le j \le n\}$ of size $m \times n$ is processed, where l_{ij} denotes the gray level of the image I at (i, j). The maximum value of the gray intensity of the image I is denoted as L. In order to consider the contextual spatial information, a spatial correlation between surrounding pixels is calculated. For this purpose, a neighborhood N of order d at given position (i, j) is used $N_{ij}^d = \{(i + u, j + v), (u, v) \in N^d\}$. The value of d determines the configuration that the neighborhood system takes [1]. This article considers the second-order neighborhood system N^2. This system can be defined in spatial terms as $(u, v) \in \{(\pm 1, 0), (0, \pm 1)(1, \pm 1)(-1, \pm 1)\}$ and is shown in Fig. 15.1.

The energy of the image I at gray intensity value $l(0 \le l \le L)$ is calculated generating a two-dimensional matrix for every intensity value as $B_l = \{b_{i,j}, 1 \le i \le m, 1 \le j \le n\}$ where $b_{i,j} = 1$ if the intensity at the current position

Fig. 15.1 Second-order neighborhood system N^2

$(i-1, j-1)$	$(i-1, j)$	$(i-1, j+1)$
$(i, j-1)$	(i, j)	$(i, j+1)$
$(i+1, j-1)$	$(i+1, j)$	$(i+1, j+1)$

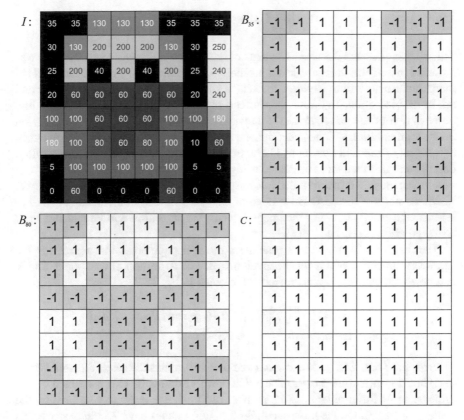

Fig. 15.2 Example image I, binary matrix B_l for $l = 35$ and $l = 80$ and constant matrix C

is greater than l the intensity value ($l_{i,j} > l$), or else $b_{i,j} = -1$. Figure 15.2 depicts a grayscale image I with the intensity values l on top of each pixel and two examples of the binary matrix B_l.

Let $C = \{c_{ij}, 1 \leq i \leq m, 1 \leq j \leq n\}$ be a constant matrix where $c_{ij} = 1, \forall(i, j)$, the energy value E_l of the image I at gray intensity value l is computed as:

$$E_l = -\sum_{i=1}^{m}\sum_{j=1}^{n}\sum_{pq \in N_{ij}^2} b_{ij} \cdot b_{pq} + \sum_{i=1}^{m}\sum_{j=1}^{n}\sum_{pq \in N_{ij}^2} c_{ij} \cdot c_{pq} \tag{15.1}$$

The right side of the summation in Eq. 15.1 is a constant term devoted to assuring a positive energy value $E_l \geq 0$. A quick look of Eq. 15.1. shows that for a given image I at intensity value l will be zero if all the elements of the binary image B_l are either 1 or -1. This approach determinates the energy associated to every intensity value of the image to generate a curve considering spatial contextual information of the image.

15.3   Image Segmentation Using ALO

This section presents the basic elements of the antlion optimizer (ALO) how it was modified to perform the multi-level thresholding over the energy curve. The selected metaheuristic algorithm explores the search space defined by the energy curve and not by the histogram. To evaluate the effects of the energy curve on the thresholding process, two of the parametric criteria widely used in the scientific community on image segmentation will be used: the one proposed by Otsu [11] based on variance and the one proposed by Kapur [12] based on entropy. Thus, the ALO algorithm described in this chapter is evaluated using the methods of Otsu and Kapur.

First, an image I is incorporated into the process to later determine its energy curve. Then, the steps of the ALO are applied to find the best threshold values required to segment the energy curve of the image I. Thus, the algorithm will be repeated until a stop criterion is reached.

The first step in ALO is to generate a random population that represents the candidate threshold values for the given image. The definition of this population $\mathbf{X}$ and how each element is constructed is presented in Eq. 15.2.

$$\mathbf{X} = [\mathbf{TH}_1, \mathbf{TH}_2, \ldots, \mathbf{TH}_N], \quad \mathbf{TH}_i = [th_1, th_2, \ldots, th_{nt}]^T,$$
$$\text{subject to} \quad th_1 < th_2 < \ldots < th_k < L \tag{15.2}$$

From Eq. 15.2, $\mathbf{TH}_i \subseteq \mathbf{X}$ and it is a vector that contains the set of thresholds (th_j) that should segment the image. The amount of thresholds in $\mathbf{TH}_i$ is defined by the dimensions of the problem nt (the number of thresholds is usually defined by the designer a priori). On the other hand, 8-bit digital images have 255 intensity levels (L), it means that the search space is defined between the bounds [0, 255] on each dimension.

Moreover, similar to most of the metaheuristic algorithms, the candidate solutions are randomly generated from a feasible search space defined by the bounds $[g_{min}, g_{max}]$, where g_{min} and g_{max} are the minimum and the maximum gray levels in the energy curve, respectively. Therefore, th_i of the i-th solution (TH) can be generated using Eq. 15.3:

$$th_i = g_{min} + rand \times (g_{max} - g_{min}), \quad \forall i = [1, 2, \ldots, K] \tag{15.3}$$

where $rand$ is a random number uniformly distributed between 0 and 1. th_i is the i-th threshold value from a feasible solution $\mathbf{TH}_i$ i corresponds to a dimension of the search space.

After that, the operators used by ALO update the current solutions of the population, then the terminal conditions are checked. If they are met, the algorithm stops and returns the best set of thresholds; otherwise, it starts a new generation. The following subsections explain the main components of the proposed approach.

15.3.1 Antlion Optimizer for Image Thresholding

The objective of this subsection is to describe the ant-lion optimization algorithm (ALO) applied to the determination of multi-level segmentation. The ALO algorithm is inspired by the behavior of an insect called ant-lion. Such insects have peculiar hunting behaviors, since they create traps in the form of wells where they wait until their prey (normal ants) falls. To make it easier for the prey to fall into the trap, ant-lions throw sand at the prey so that it slides and can be caught.

From a computational point of view, an original population of $\mathbf{X}$ of N individuals is created, and then the quality of the thresholds proposed by each particle is evaluated using the Otsu or Kapur as objective function. Then, in each iteration the ALO algorithm modifies the position of every individual according to a random walk described by Eq. 15.4.

$$\mathbf{TH}_i^t = \frac{\left(\mathbf{TH}_i^t - a_i\right) \times \left(d_i - c_i^t\right)}{\left(d_i^t - a_i\right)} + c_i \tag{15.4}$$

where c_i^t y d_i^t are the minimum and maximum values of the i-th design variable in iteration t. Equation 15.4 represents the normalized random walk restricted to the limits of the search space.

The next step in the ALO algorithm determines how the random walk is affected by ant-lion traps. This behavior is described by the following expression,

$$c_i^t = Antlion_j^t + c^t$$
$$d_i^t = Antlion_j^t + d^t \tag{15.5}$$

where c^t and d^t are the minimum and maximum of all variables at t iteration. Meanwhile, the roulette wheel operator is used to select the j-th $Antlion_j^t$ at the t iteration. This element is a candidate solution $\mathbf{TH}_i$ extracted from $\mathbf{X}$. Equation 15.5 establishes that the random walk of ants in a hypersphere defined by the vectors c and d around a selected antlion. Based on this concept, ant-lions are restricted to move within the space delimited by the hyper-sphere that has a prey in the centre. The ALO algorithm builds the ant-hunting strategy using the roulette operator, which is used for the selection of an ant-lion (since ants fall only into a ant-lion trap) based on the values of the aptitude function. Following this strategy, ant-lions with the best fitness value are more likely to catch common ants.

At the end of the trap construction stage, common ants slide down to fall next to a ant-lion, which throws sand from the center of the well to make the ant fall faster. To mimic this behavior, the radius of each hyper-sphere is diminished adaptively using the following equation:

$$c^t = \frac{c^t}{rad}, \; d^t = \frac{d^t}{rad}, \; rad = \frac{10^w t}{T} \tag{15.6}$$

rad is a radius for exploitation, w is a constant defined based on the current iteration ($w = 2$ when $t < 0.1T$, $w = 3$ when $t < 0.5T$, $w = 4$ when $t < 0.75T$ $w = 5$ when $t < 0.9T$ and $w = 10$ when $t < 0.95T$). In other words, w adjusts the intensity of exploitation. Then, in the final stage, the objective is to catch the prey when it reaches the bottom of the well. In this step, the ant-lion is removed so that the ant can be captured. To formulate this process mathematically the following equation is used:

$$Antlion_j^t = Ant_i^t \text{ if } f\left(Ant_i^t\right) > f\left(Antlion_j^t\right) \tag{15.7}$$

From Eq. 15.7 *Antlion*$_j^t$ and *Ant*$_j^t$ are the positions ($\mathbf{TH}_i$) selected from $\mathbf{X}$ at the t iteration. This equation considers that whenever the ants become fitter than its corresponding antlion, the prey catching occurs. Then the antlion updates its position to the latest position of the hunted ant to enhance its chance of catching new preys.

The concept of elitism is applied in this algorithm to maintain the best solution obtained in each iteration and later affect the movements of the rest of ants. In this sense, each ant walks randomly around a selected ant-lion (through the roulette operator) and elite ant simultaneously using Eq. 15.8.

$$Ant_i^t = \frac{R_A^t + R_E^t}{2} \tag{15.8}$$

From Eq. 15.8 R_A^t and R_E^t are random walks around the selected lion ant and elite lion ant in each iteration t respectively. Based on all previous steps of ALO, the two random walks are used to update the position of the ants. In addition, parameters c and d are updated with respect to the current iteration according to Eq. 15.5. The aptitude function is used to determine the prey ants and based on this evaluation the ant-lions update their position according to the following: if any of the prey ants turns out to be more suitable than any other lion ant, then the position of the lion ant is moved to the position of the prey ant in the following equation. Then, the best lion ant (elite) is updated. The above steps are repeated until a previously established stop criterion is reached.

As previously mentioned, the ALO algorithm attempts to mimic the hunting mechanisms of the ant-lion. Unlike other methods based on ant behaviors such as ACO or ACOR, in the ALO algorithm the ant-lion is a predator that generates a trap for ants to fall prey. In addition, in the ACO algorithm the metaphor involves searching for the best food sources while reinforcing the trajectory generated by the use of pheromones.

15.4   Experimental Results

The ALO algorithm applied to the multi-level thresholding of images is evaluated on a set of test images containing different levels of complexity. Figure 15.3 shows these five images together with their histograms and energy curves. For visual com-

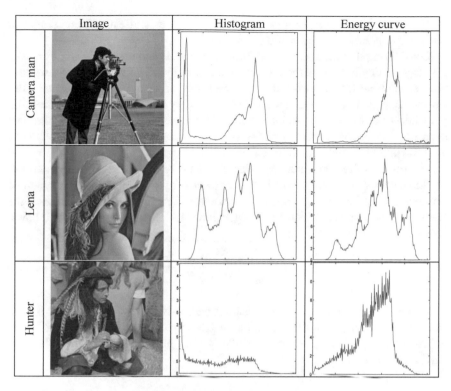

Fig. 15.3 Example images and their histogram and energy curve

parisons only three significant and representative examples have been selected from the total set to illustrate graphical and numerical results. It should be noted that all the images shown have different characteristics and distributions. In this context, the energy curve preserves some characteristics related to the mentioned distribution, incorporating also the contextual information of each pixel. A clear example of the differences between the histogram of an image and the energy curve can be clearly seen in the Hunter image, where the histogram presents a flat tendency distribution after a peak in the low intensity zone, while the energy curve shows a significant mode in the central zone of the range of values. In the other two images the differences are not so evident showing in both cases different modes.

In order to verify the effectiveness of the proposed approaches, they have been compared against different metaheuristic algorithms that are part of classical methods and state of the art. Among the algorithms used are particle swarm optimization (PSO) [13], crow search algorithm (CSA) [14], runner root algorithm (RRA) [15] and ant colony optimization for continuous domains (ACOR) [16]. All the selected metaheuristic algorithms are based on the recommendations proposed by their authors. According to the NFL theorem, not all optimization algorithms can be efficiently applied to the same problem [17]. In this context, experiments and comparisons between metaheuristic algorithms applied to the threshold can help determine if ALO is competitive for real-world applications.

Considering that metaheuristic algorithms contain random variables, it is necessary to statistically analyze their results. For this purpose, 35 independent tests have been carried out for each experiment carried out. Segmentation is performed considering 2, 4, 8, 16 and 32 nt thresholds on the energy curve. The stop criterion and population size for each algorithm is 500 and 50 respectively for each algorithm. All experiments were performed using MATLAB 8.3 on an Intel Xeon-2620 v3 CPU @ 2.4 GHz processor with 16 GB of RAM.

The results of the thresholding based on ALO are measured not only considering the objective function values but also considering the quality of the segmented images. One of the metrics used is the Standard Deviation (σ) that represents the stability of the results obtained by the algorithms [18]. The σ is computed as follows:

$$\sigma = \sqrt{\sum_{i=1}^{Iter\max} \frac{(\sigma_i - \mu)}{Ru}} \tag{15.9}$$

Another interesting metric is the Peak-Signal-to-Noise Ratio (PSNR). It is used compare the similarity of the segmented image with the original image. The PSNR is based on the mean square error (MSE) of each pixel [19–22]. Both PSNR and RMSE are defined as:

$$PSNR = 20 \log_{10}\left(\frac{255}{RMSE}\right), \text{(dB)}$$

$$RMSE = \sqrt{\frac{\sum_{i=1}^{ro}\sum_{j=1}^{co}(I_{or}(i,j) - I_{th}(i,j))}{ro \times co}} \tag{15.10}$$

From Eq. 15.10 I_{or} is the original image, I_{th} is the segmented image. Meanwhile, ro and co are the maximum number of rows and columns of the image. In the same context, the Structure Similarity Index (SSIM) compares the structures of the original segmented image [23], and it is defined in Eq. 15.11. A higher SSIM value represents a better segmentation of the original image.

$$SSIM\,(I_{or}, I_{th}) = \frac{\left(2\mu_{I_{or}}\mu_{I_{th}} + C1\right)\left(2\sigma_{I_{or}I_{th}} + C2\right)}{\left(\mu_{I_{or}}^2 + \mu_{I_{th}}^2 + C1\right)\left(\sigma_{I_{or}}^2 + \sigma_{I_{th}}^2 + C2\right)}$$

$$\sigma_{I_{th}I_{or}} = \frac{1}{N-1}\sum_{i=1}^{N}\left(I_{or_i} + \mu_{I_{or}}\right)\left(I_{th_i} + \mu_{I_{th}}\right) \tag{15.11}$$

In Eq. 15.11 the mean of the original image is $\mu_{I_{or}}$ and the mean of the thresholded image is represented by $\mu_{I_{th}}$. In the same way, for each image, the values of $\sigma_{I_{Gr}}$ and $\sigma_{I_{th}}$ correspond to the standard deviation. $C1$ and $C2$ are constants used to avoid the instability when $\mu_{I_{Gr}}^2 + \mu_{I_{th}}^2 \approx 0$. The values of $C1$ and $C2$ are set to 0.065 considering the experiments of [21].

In the same context, the Feature Similarity Index (FSIM) [24], calculates the similarity between two images: in this case, the original grayscale image, and the segmented image. As PSNR and SSIM the higher value is interpreted as better performance of the thresholding method. The FSIM is then defined as:

$$FSIM = \frac{\sum\limits_{w \in \Omega} S_L(w)PC_m(w)}{\sum\limits_{w \in \Omega} PC_m(w)} \tag{15.12}$$

On the FSIM the entire domain of the image is defined by Ω, and their values are computed by Eq. 15.13.

$$\begin{aligned} S_L(w) &= S_{PC}(w)S_G(w) \\ S_{PC}(w) &= \frac{2PC_1(w)PC_2(w)+T_1}{PC_1^2(w)+PC_2^2(w)+T_1} \\ S_G(w) &= \frac{2G_1(w)G_2(w)+T_2}{G_1^2(w)+G_2^2(w)+T_2} \end{aligned} \tag{15.13}$$

G is the gradient magnitude (GM) of a digital image and is defined, and the value of PC that is the phase congruence is defined as follows:

$$G = \sqrt{G_x^2 + G_y^2}$$

$$PC(w) = \frac{E(w)}{\left(\varepsilon + \sum\limits_n A_n(w)\right)} \tag{15.14}$$

where $A_n(w)$ is the local amplitude on scale n and $E(w)$ is the magnitude of the response vector in w on n. ε is a small positive number and $PC_m(w) = \max(PC_1(w), PC_2(w))$.

15.4.1 Results Using Otsu's Method

This subsection presents the results obtained by the ALO algorithm using Otsu's [11] formulation. For this ALO uses the formulation of Otsu as objective function in order to maximize the variance between classes. The results are obtained on five images with five different threshold numbers. Table 15.1 presents a table with the best thresholds obtained according to their aptitude value by the different threshold methods. In Table 15.1, column I corresponds to the name of the image, column nt indicates the number of thresholds used ($nt = 2, 4, 8, 16, 32$). The rest of the columns correspond to the thresholds obtained by the algorithms ALO, PSO, CSA, RRA and

ACOR. These thresholds can be applied directly to the energy curve to segment the image.

The objective of multi-level segmentation is to obtain the best possible quality in the segmented images through the use of thresholds. In this context, the PSNR results between the segmented image and the original image should be analyzed. To statistically compare the solutions obtained by the different approaches, the mean of the proficiency value generated by the target function and its standard deviation is calculated (σ). The values corresponding to this analysis are shown in Table 15.2.

In Table 15.2, a higher PSNR mean value represents better segmentation. In addition, a lower value of σ is expected to reflect less variability among the results obtained by each experiment. A similar study is presented in Table 15.3 where the behavior of the SSIM metric described in the previous section is evaluated, which aims to analyze how image structures are affected by segmentation, a higher value indicates better segmentation.

Another interesting metric that has been used to verify the quality of a segmented image is FSIM described in the previous section. Table 15.4 shows the FSIM results using the results obtained by each of the selected metaheuristic algorithms. The FSIM metric indicates a higher quality when it reaches a higher value.

In summary, the metrics presented in Tables 15.2, 15.3 and 15.4 are used to analyze the quality of the segmented images using the best thresholds found by each of the methods compared. The PSNR, SSIM and FSIM values observed indicate that the ALO algorithm may find better thresholds compared to the other methodologies analyzed.

Figure 15.4 shows the images obtained by the ALO algorithm with different numbers of thresholds using Otsu as an objective function. Specifically, Fig. 15.4 is based on the best thresholds found by the ALO algorithm. The resulting images show how the increase in the number of thresholds significantly increases the quality of the images, especially from a qualitative point of view. Note in this sense the images with $nt = 32$.

15.4.2   Results Using Maximum Entropy

This subsection presents the results of the algorithms selected to perform multilevel threshold using as an objective function the entropy proposed by Kapur [12]. The experimentation process is the same as in the previous subsection. Table 15.5 presents a comparison between the thresholds obtained by each of the selected algorithms (ALO, PSO, CSA, RRA and ACOR).

The thresholds obtained by each metaheuristic algorithm are applied to the energy curve of each image to classify the pixels. When the classification is carried out, a segmented image is generated as a result, in such a way that different quality metrics are evaluated in order to proceed to its analysis. Similar to the previous sections, the quality of the segmentation is measured in terms of PSNR, SSIM and FSIM.

Table 15.1 Thresholds obtained by ALO, PSO, CSA, RRA and ACOR algorithms using the Otsu method as an objective function

I	nt	ALO	PSO	CSA	RRA	ACOR
Cameraman	2	83 153	87 151	83 153	83 153	83 153
	4	78 141 168 202	70 143 159 177	78 141 168 202	78 141 168 202	72 141 165 195
	8	40 86 120 140 156 168 179 207	47 81 126 158 177 204 234 236	38 84 118 139 156 168 179 208	38 84 118 139 156 168 179 208	37 78 109 128 144 160 175 205
	16	26 51 78 100 114 125 135 145 154 161 168 175 182 193 211 230	21 41 42 45 66 90 99 114 132 139 155 164 175 192 198 204	24 53 80 104 119 131 141 150 158 165 171 178 184 193 208 228	24 53 80 104 119 131 141 150 158 165 171 178 184 193 208 228	23 39 56 76 97 111 123 133 143 152 160 168 176 183 197 222
	32	15 23 38 52 65 75 85 97 105 113 121 127 133 139 145 150 154 157 161 164 168 171 175 179 183 187 196 206 216 229 236 242	31 39 43 46 46 56 57 61 76 85 93 96 103 105 113 118 132 138 140 147 158 162 166 167 174 175 180 188 209 211 234 255	14 23 29 35 41 56 72 82 92 105 113 120 127 133 140 146 153 158 162 166 170 174 177 179 183 185 188 193 205 215 222 235	14 23 29 35 41 56 72 82 92 105 113 120 127 133 140 146 153 158 162 166 170 174 177 179 183 185 188 193 205 215 222 235	1 1 12 16 23 32 42 52 62 73 85 96 105 112 119 126 133 140 146 153 158 163 167 172 177 181 185 191 199 209 221 235
Lena	2	114 169	115 168	114 169	114 169	114 169
	4	80 119 149 183	72 100 135 186	80 119 149 183	80 119 149 183	80 119 149 183
	8	67 92 114 134 150 166 184 203	3 51 97 120 150 163 178 203	63 90 113 133 149 165 184 203	63 90 113 133 149 165 184 203	63 86 105 121 137 153 172 195

(continued)

Table 15.1 (continued)

I	nt	ALO	PSO	CSA	RRA	ACOR
	16	48 58 72 85 97 109 121 132 142 151 160 170 182 193 203 213	3 23 54 57 72 91 106 120 133 149 169 174 183 197 239 246	50 61 74 90 104 116 126 135 145 154 167 181 195 205 215 243	50 61 74 90 104 116 126 135 145 154 167 181 195 205 215 243	46 55 65 78 90 101 111 122 132 142 151 160 171 184 197 209
	32	7 36 46 51 56 66 74 83 90 95 101 105 111 116 122 127 133 139 145 150 155 160 167 174 181 189 195 200 205 210 217 241	5 20 25 26 29 39 44 48 52 53 53 58 63 69 79 97 103 114 115 123 133 136 149 151 162 177 181 190 201 209 215 251	3 28 31 49 56 64 70 75 81 87 96 104 111 118 124 129 133 140 146 151 156 160 165 170 175 181 187 195 202 208 215 226	3 28 31 49 56 64 70 75 81 87 96 104 111 118 124 129 133 140 146 151 156 160 165 170 175 181 187 195 202 208 215 226	1 1 37 43 47 51 55 61 67 75 83 90 97 104 111 119 126 132 139 145 151 156 161 167 173 178 184 191 197 203 209 217
Hunter	2	87 141	87 142	87 141	87 141	87 141
	4	56 96 130 160	53 106 131 154	56 96 130 160	56 96 130 160	56 96 130 160
	8	38 65 89 110 128 145 161 184	54 79 114 128 152 187 214 247	42 71 97 118 135 151 166 187	42 71 97 118 135 151 166 187	34 60 83 105 124 142 160 183
	16	20 36 51 65 79 93 105 116 125 134 143 152 160 169 181 199	10 18 18 42 53 70 73 93 115 122 129 141 156 176 187 230	24 44 60 75 88 100 109 117 124 131 141 151 160 170 183 201	24 44 60 75 88 100 109 117 124 131 141 151 160 170 183 201	18 33 46 59 72 85 99 110 121 131 141 151 160 169 182 200

(continued)

Table 15.1 (continued)

I	nt	ALO	PSO	CSA	RRA	ACOR
	32	1 5 12 20 30 41 48 56 64 71 78 86 93 100 105 111 117 123 129 135 141 146 152 158 162 167 173 180 189 199 208 221	28 41 56 66 83 102 116 119 125 128 130 132 139 144 153 154 160 166 181 192 196 197 202 206 212 213 228 234 235 236 237 240	8 14 20 31 40 49 57 64 71 77 83 89 97 105 110 115 120 126 131 139 148 156 165 170 176 184 192 200 207 214 223 234	8 14 20 31 40 49 57 64 71 77 83 89 97 105 110 115 120 126 131 139 148 156 165 170 176 184 192 200 207 214 223 234	1 9 17 24 33 41 49 58 66 74 82 90 98 106 113 119 125 131 137 142 147 152 157 162 167 172 177 184 191 199 209 221
Peppers	2	122 172	119 170	122 172	122 172	122 172
	4	79 120 154 182	86 133 173 197	79 120 154 182	79 120 154 182	79 119 153 181
	8	54 86 109 131 151 168 184 201	34 70 108 128 151 176 199 243	52 85 109 132 152 169 184 201	52 85 109 132 152 169 184 201	50 79 98 115 137 157 178 198
	16	36 56 74 88 98 108 119 131 142 153 162 170 179 188 197 207	38 55 71 77 87 89 102 128 150 160 164 175 181 185 205 226	34 53 70 81 91 99 109 121 135 148 159 169 178 187 195 204	34 53 70 81 91 99 109 121 135 148 159 169 178 187 195 204	29 45 60 72 84 94 103 114 127 139 150 161 171 181 191 203
	32	11 22 35 46 55 65 73 81 89 96 102 109 116 122 128 134 140 145 150 155 160 164 169 175 180 185 189 192 197 204 211 242	14 14 31 38 48 58 61 63 64 69 72 76 85 97 112 112 119 131 139 149 155 163 165 174 183 187 202 204 207 211 226 255	6 22 33 43 52 69 82 92 99 104 109 113 118 121 123 125 129 136 144 152 159 165 170 176 180 186 191 197 203 209 214 223	6 22 33 43 52 69 82 92 99 104 109 113 118 121 123 125 129 136 144 152 159 165 170 176 180 186 191 197 203 209 214 223	15 20 24 31 37 46 55 64 72 79 87 94 100 106 112 120 128 136 143 150 156 163 169 175 180 185 189 193 198 205 212 229

(continued)

Table 15.1 (continued)

I	nt	ALO	PSO	CSA	RRA	ACOR
Butterfly	2	110 159	109 158	110 159	110 159	110 159
	4	85 118 150 180	95 136 152 179	89 122 153 179	89 122 153 179	85 118 150 179
	8	72 93 110 126 142 159 179 195	78 100 113 133 147 159 184 240	50 74 94 110 126 142 162 184	50 74 94 110 126 142 162 184	66 82 96 111 126 142 160 182
	16	28 55 68 81 94 102 111 118 127 137 146 158 170 180 189 203	2 7 13 40 70 90 102 114 116 129 141 158 166 176 182 192	50 64 75 86 99 110 120 130 142 150 160 171 184 195 210 249	50 64 75 86 99 110 120 130 142 150 160 171 184 195 210 249	45 57 65 75 84 93 102 113 125 134 145 158 169 179 191 201
	32	11 35 48 55 67 75 81 89 95 99 101 110 116 123 131 137 141 147 151 154 157 159 166 170 177 184 188 191 195 198 211 234	1 34 58 66 67 79 86 92 100 105 108 115 119 119 123 134 144 157 163 165 168 171 183 189 190 193 194 199 207 208 217 223	29 39 46 51 56 61 71 82 89 94 99 102 106 111 119 126 131 137 142 149 158 167 170 174 177 187 195 203 211 215 216 237	29 39 46 51 56 61 71 82 89 94 99 102 106 111 119 126 131 137 142 149 158 167 170 174 177 187 195 203 211 215 216 237	1 31 40 47 50 56 62 66 72 76 82 89 96 102 109 115 121 127 135 142 146 151 154 162 170 174 179 186 190 194 201 207

Table 15.2 Mean and σ of the PSNR values obtained by the ALO, PSO, CSA, RRA and ACOR algorithms using Otsu as the target function

I	nt	ALO		PSO		CSA		RRA		ACOR	
		Mean	σ	Mean	σ	Mean	σ	Mean	σ	Mean	σ
Cameraman	2	**17.6057**	7.20E−15	**19.7264**	1.20E−01	13.8636	7.20E−15	**17.6057**	7.20E−15	17.3993	1.55E−01
	4	19.4286	7.20E−15	**19.7264**	8.10E−01	19.1721	1.50E−02	19.4246	1.80E−02	19.4852	7.11E−01
	8	25.4905	2.50E−02	25.0233	9.10E−01	25.0992	2.10E−01	**25.5142**	3.10E−01	25.5398	1.49E−01
	16	**29.7947**	3.60E−01	28.7481	7.90E−01	29.5755	2.40E−01	29.4064	5.40E−01	29.6193	5.92E−02
	32	35.5253	1.60E+00	32.0634	1.50E+00	**37.4424**	1.50E+00	34.3886	2.00E+00	35.6078	4.43E−01
Lena	2	**13.3913**	3.60E−15	13.3956	4.00E−02	10.5754	3.60E−15	13.3913	3.60E−15	13.1654	5.26E−02
	4	18.3293	7.20E−15	18.3399	2.30E−01	16.8339	1.80E−03	**18.3294**	4.20E−04	17.9957	5.73E−02
	8	20.9123	1.50E−01	20.7929	1.10E+00	20.7413	9.60E−02	**21.1996**	9.90E−01	20.8341	1.29E−01
	16	**26.8999**	2.20E+00	25.0613	2.80E+00	27.0567	1.10E+00	25.5768	2.60E+00	26.2025	5.43E−01
	32	34.9573	1.90E+00	30.633	3.60E+00	**37.3675**	1.90E+00	33.8087	2.70E+00	33.3224	5.87E−01
Hunter	2	**17.1166**	1.10E−14	17.1105	3.20E−02	14.2921	1.10E−14	17.1166	1.10E−14	16.8552	8.47E−02
	4	21.3132	7.20E−15	21.0799	5.90E−01	18.752	2.00E−02	**21.3332**	4.60E−02	20.9367	5.61E−02
	8	**25.6913**	1.50E−01	24.4137	9.70E−01	24.3898	2.40E−01	25.6086	5.10E−01	25.5616	1.16E−01
	16	**31.1252**	3.10E−01	28.902	9.80E−01	30.9959	6.80E−01	30.7786	7.80E−01	31.068	1.48E−01
	32	36.6525	4.80E−01	32.9076	1.20E+00	**37.0924**	7.20E−01	35.8073	8.70E−01	36.5639	1.56E−01
Peppers	2	12.2336	5.40E−15	**12.2582**	4.10E−02	11.251	5.40E−15	12.2337	5.40E−15	12.0117	2.90E−02
	4	**19.0335**	3.60E−15	19.0131	3.90E−01	18.0765	1.10E−02	19.033	2.20E−02	18.7062	5.32E−02
	8	**24.2045**	8.60E−02	23.9479	5.70E−01	23.7142	1.80E−01	23.3095	2.60E−01	24.1973	3.62E−02
	16	29.6949	5.50E−01	27.8284	1.30E+00	29.8357	7.00E−01	29.4878	9.00E−01	**29.9106**	2.36E−01

(continued)

Table 15.2 (continued)

I	nt	ALO		PSO		CSA		RRA		ACOR	
		Mean	σ	Mean	σ	Mean	σ	Mean	σ	Mean	σ
	32	**35.9873**	7.00E−01	32.5848	1.60E+00	34.464	9.90E−01	35.4113	9.90E−01	35.7821	2.80E−01
Butterfly	2	12.5859	5.40E−15	**12.5951**	1.10E−02	10.2748	5.40E−15	12.586	5.40E−15	12.3557	3.68E−02
	4	**16.7709**	1.90E−02	16.7617	7.00E−01	15.5824	1.00E−01	16.7221	4.10E−01	16.564	2.24E−01
	8	20.6202	8.20E−01	21.0632	1.70E+00	20.335	3.70E−01	**22.2657**	1.90E+00	20.9489	4.08E−01
	16	**29.4413**	1.50E+00	26.136	2.60E+00	28.8109	1.30E+00	28.6761	2.40E+00	28.9924	1.36E+00
	32	35.4651	9.90E−01	32.5002	3.10E+00	**37.734**	1.60E+00	35.1612	1.60E+00	36.1365	1.20E+00

Table 15.3 Mean and σ of the SSIM values obtained by the ALO, PSO, CSA RRA and ACOR algorithms using Otsu as the target function

	nt	ALO		PSO		CSA		RRA		ACOR	
		Mean	σ	Mean	σ	Mean	σ	Mean	σ	Mean	σ
Cameraman	2	**0.7544**	3.40E−16	0.7542	2.30E−03	0.5263	3.40E−16	**0.7544**	3.40E−16	0.7401	2.67E−03
	4	0.7789	1.10E−16	**0.7858**	1.60E−02	0.6212	2.10E−04	0.7789	3.40E−04	0.7706	1.21E−02
	8	**0.8808**	4.20E−04	0.8767	1.80E−02	0.7439	2.80E−03	**0.8808**	3.70E−03	0.87	2.46E−03
	16	**0.9236**	9.20E−03	0.9177	1.70E−02	0.8218	5.60E−03	0.9196	8.30E−03	0.906	6.42E−04
	32	**0.9768**	1.70E−02	0.9465	2.30E−02	0.9719	1.50E−02	0.9644	2.30E−02	0.9625	3.22E−03
Lena	2	**0.5439**	0.00E+00	0.5442	2.80E−03	0.2945	0.00E+00	**0.5439**	0.00E+00	0.5325	2.06E−03
	4	**0.7475**	4.50E−16	0.7475	9.90E−03	0.596	3.10E−04	**0.7475**	9.90E−05	0.733	1.09E−03
	8	0.8311	4.60E−03	0.8225	2.80E−02	0.7522	2.90E−03	**0.8362**	2.20E−02	0.823	3.55E−03
	16	**0.9351**	2.40E−02	0.9058	3.90E−02	0.8951	1.30E−02	0.9174	3.10E−02	0.914	5.72E−03
	32	**0.9858**	7.70E−03	0.9613	2.60E−02	0.9799	1.10E−02	0.9804	1.40E−02	0.9651	2.20E−03
Hunter	2	**0.6558**	2.30E−16	0.6554	1.80E−03	0.3002	2.30E−16	**0.6558**	2.30E−16	0.644	1.30E−03
	4	0.8082	3.40E−16	0.8028	2.50E−02	0.4687	8.20E−04	**0.8092**	2.20E−03	0.7928	1.01E−03
	8	**0.9099**	3.60E−03	0.883	2.90E−02	0.6712	6.20E−03	0.9094	1.30E−02	0.9015	3.30E−03
	16	**0.9713**	2.80E−03	0.9528	1.60E−02	0.8469	7.40E−03	0.9688	8.60E−03	0.9564	1.62E−03
	32	**0.9929**	2.00E−03	0.9779	8.30E−03	0.9571	3.00E−03	0.9889	4.20E−03	0.9745	6.53E−04
Peppers	2	0.4843	2.30E−16	**0.4854**	2.30E−03	0.3272	2.30E−16	0.4843	2.30E−16	0.4749	1.05E−03
	4	**0.7918**	5.60E−16	0.7915	1.40E−02	0.623	4.60E−04	0.7918	9.80E−04	0.7774	1.44E−03
	8	0.9065	1.30E−03	0.9032	1.40E−02	0.762	1.60E−03	**0.9082**	3.80E−03	0.8938	9.16E−04
	16	**0.9653**	5.40E−03	0.9483	1.70E−02	0.8814	6.70E−03	0.9633	9.40E−03	0.9519	2.26E−03

(continued)

Table 15.3 (continued)

	nt	ALO		PSO		CSA		RRA		ACOR	
		Mean	σ	Mean	σ	Mean	σ	Mean	σ	Mean	σ
	32	**0.9934**	2.20E−03	0.979	1.20E−02	0.9609	3.90E−03	0.991	3.80E−03	0.9739	8.27E−04
Butterfly	2	0.5234	1.10E−16	**0.5235**	9.40E−04	0.2238	1.10E−16	0.5234	1.10E−16	0.5132	5.56E−04
	4	**0.7473**	2.10E−04	0.7446	3.20E−02	0.5152	3.90E−03	0.7446	1.70E−02	0.737	8.55E−03
	8	0.8736	1.80E−02	0.8791	3.50E−02	0.7324	9.00E−03	**0.905**	3.60E−02	0.8736	8.26E−03
	16	**0.9809**	8.80E−03	0.9512	2.80E−02	0.9116	8.60E−03	0.9732	2.20E−02	0.9613	8.08E−03
	32	**0.9957**	1.80E−03	0.9855	1.50E−02	0.9791	2.90E−03	0.9944	3.40E−03	0.9766	1.69E−03

Table 15.4 Mean and σ of FSIM values obtained by ALO, GA, PSO, CSA, RRA and ACOR algorithms using Otsu as objective function

I	nt	ALO Mean	ALO σ	PSO Mean	PSO σ	CSA Mean	CSA σ	RRA Mean	RRA σ	ACOR Mean	ACOR σ
Cameraman	2	**0.7479**	3.40E−16	0.7478	2.30E−03	0.6749	5.60E−16	**0.7479**	5.60E−16	0.7358	2.80E−03
	4	0.7822	1.10E−16	**0.7926**	1.60E−02	0.7824	1.50E−04	0.7822	1.10E−04	0.7807	2.02E−02
	8	0.9158	4.20E−04	0.906	1.80E−02	0.9105	2.90E−03	**0.9164**	3.90E−03	0.9056	2.88E−03
	16	**0.9682**	9.20E−03	0.953	1.70E−02	0.9668	2.40E−03	0.9638	5.80E−03	0.9524	4.28E−04
	32	0.9838	1.70E−02	0.9732	2.30E−02	**0.9873**	1.90E−03	0.9825	3.10E−03	0.9637	5.81E−04
Lena	2	**0.6609**	0.00E+00	0.6609	2.80E−03	0.5961	2.30E−16	**0.6609**	2.30E−16	0.6499	3.44E−03
	4	**0.7974**	4.50E−16	0.7959	9.90E−03	0.7391	8.50E−05	**0.7974**	3.30E−05	0.7825	2.28E−03
	8	0.8757	4.60E−03	0.8617	2.80E−02	0.8668	2.20E−03	**0.8761**	7.40E−03	0.8607	1.47E−03
	16	0.933	2.40E−02	0.9112	3.90E−02	**0.9333**	2.80E−03	0.9225	1.50E−02	0.9024	3.72E−03
	32	0.9802	7.70E−03	0.9539	2.60E−02	**0.9903**	1.20E−02	0.9741	1.60E−02	0.9596	3.07E−03
Hunter	2	**0.7615**	2.30E−16	0.7611	1.80E−03	0.6477	3.40E−16	0.7615	3.40E−16	0.7512	5.06E−03
	4	0.8759	3.40E−16	0.8711	2.50E−02	0.8119	4.10E−04	**0.8764**	1.10E−03	0.8601	2.38E−03
	8	**0.9486**	3.60E−03	0.9287	2.90E−02	0.9299	3.40E−03	0.9485	7.30E−03	0.9353	2.03E−03
	16	**0.9856**	2.80E−03	0.9722	1.60E−02	0.9852	4.30E−03	0.9838	5.10E−03	0.9684	9.15E−04
	32	**0.9968**	2.00E−03	0.9872	8.30E−03	0.9968	1.80E−03	0.9947	2.30E−03	0.9775	2.68E−04
Peppers	2	0.6921	2.30E−16	**0.6926**	2.30E−03	0.6509	3.40E−16	0.6921	3.40E−16	0.6825	6.15E−03
	4	**0.8047**	5.60E−16	0.8032	1.40E−02	0.7784	2.20E−04	0.8046	3.90E−04	0.7915	4.19E−03
	8	**0.9081**	1.30E−03	0.8955	1.40E−02	0.8951	8.50E−04	0.9072	2.20E−03	0.8918	1.11E−03
	16	**0.9656**	5.40E−03	0.9434	1.70E−02	0.9655	3.10E−03	0.9638	4.50E−03	0.9498	6.91E−04

(continued)

Table 15.4 (continued)

I	nt	ALO		PSO		CSA		RRA		ACOR	
		Mean	σ	Mean	σ	Mean	σ	Mean	σ	Mean	σ
	32	0.9911	2.20E−03	0.9739	1.20E−02	**0.9923**	2.70E−03	0.9884	3.40E−03	0.9721	6.71E−04
Butterfly	2	**0.708**	1.10E−16	0.7076	9.40E−04	0.6287	2.30E−16	0.708	2.30E−16	0.6961	3.93E−03
	4	**0.8203**	2.10E−04	0.8156	3.20E−02	0.7813	4.30E−04	0.8191	3.10E−03	0.8027	2.47E−03
	8	0.9028	1.80E−02	0.8991	3.50E−02	0.8935	3.10E−03	**0.9145**	1.70E−02	0.8904	4.12E−03
	16	**0.9734**	8.80E−03	0.9465	2.80E−02	0.9684	8.00E−03	0.9663	1.70E−02	0.9526	8.81E−03
	32	0.9926	1.80E−03	0.9792	1.50E−02	0.997	4.50E−03	0.9904	5.20E−03	0.9743	3.06E−03

Table 15.5 Thresholds obtained by ALO, PSO, CSA, RRA and ACOR algorithms using the Kapur method as objective function

I	nt	ALO	PSO	CSA	RRA	ACOR
Cameraman	2	107 196	109 198	107 196	107 196	107 196
	4	55 98 146 196	26 91 132 202	55 98 146 196	55 98 146 196	55 98 146 196
	8	22 47 72 98 126 157 193 222	26 46 60 81 107 173 202 213	22 50 78 102 129 158 193 222	22 46 69 94 121 150 193 222	23 53 83 107 134 160 193 222
	16	7 20 36 53 70 87 102 117 133 147 160 176 192 207 224 241	3 26 44 56 77 94 103 115 134 168 175 194 202 214 218 223	6 19 34 49 62 75 88 99 112 126 142 157 173 191 207 229	20 38 56 70 84 97 109 120 131 146 160 175 192 207 222 236	7 19 33 48 63 79 95 109 124 140 157 173 191 201 216 233
	32	7 17 23 34 43 50 59 67 75 84 92 102 109 115 122 129 135 141 147 154 160 166 176 183 191 200 207 215 222 230 236 244	13 18 24 28 29 31 40 44 51 52 52 55 66 74 89 96 107 112 115 127 128 141 165 167 182 192 195 207 213 216 225 234	7 13 19 25 31 37 45 52 58 65 71 78 85 94 102 110 121 132 138 147 162 169 179 189 197 201 205 211 217 225 233 242	7 19 26 33 39 44 51 60 68 76 84 91 98 107 115 123 129 136 143 149 156 163 171 181 191 201 210 218 224 229 235 243	7 12 18 24 30 37 43 50 57 64 71 79 87 95 103 112 118 124 132 141 149 157 165 173 182 191 198 205 214 224 233 243
Lena	2	94 163	89 161	94 163	94 163	94 163
	4	23 85 125 174	91 134 177 204	23 85 125 174	23 77 113 161	23 85 125 174
	8	23 58 85 112 140 169 197 245	19 82 112 132 140 160 187 212	23 62 89 114 137 159 180 202	23 58 87 113 137 159 179 201	23 60 86 111 133 154 177 201
	16	10 23 42 59 75 91 105 119 134 147 162 177 191 204 217 245	9 16 33 46 48 73 93 99 110 120 151 156 159 173 190 211	23 39 54 67 81 94 109 125 139 152 166 179 192 208 224 245	11 23 33 46 63 81 96 108 121 134 147 161 177 194 212 224	11 23 37 51 64 78 93 107 121 135 149 163 177 190 203 217

(continued)

Table 15.5 (continued)

I	nt	ALO	PSO	CSA	RRA	ACOR
	32	4 10 16 23 33 40 47 54 61 69 77 88 95 102 107 114 119 125 131 135 142 153 158 165 173 182 194 206 218 223 230 245	2 6 17 17 28 42 45 49 56 71 81 95 99 106 117 118 129 146 152 167 173 179 189 201 204 204 206 219 223 236 244 247	13 20 32 38 44 49 55 63 70 77 83 93 103 112 122 129 136 143 149 157 166 177 188 201 213 224 238 241 244 245 249 252	5 11 17 23 27 32 37 43 49 55 60 66 72 78 84 90 97 104 111 127 144 159 171 179 188 194 201 208 215 223 230 245	2 6 12 17 23 33 39 45 53 61 69 77 86 94 102 109 117 126 135 143 151 160 168 175 183 190 198 206 213 219 230 245
Hunter	2	93 179	99 179	93 179	93 179	93 179
	4	45 93 133 179	32 82 114 177	45 93 133 179	41 89 131 179	45 93 133 179
	8	29 58 89 117 145 174 198 227	16 45 100 153 181 205 218 227	25 49 73 98 125 152 182 222	23 52 81 109 140 175 197 227	25 49 75 101 127 154 183 222
	16	17 32 46 61 76 90 104 118 132 148 163 178 192 206 220 233	16 26 30 52 74 81 111 134 140 162 182 196 201 203 218 222	13 26 41 57 73 88 103 118 133 148 163 178 192 206 221 233	18 35 53 69 85 100 115 130 144 157 171 182 194 207 220 234	13 24 37 51 65 81 96 111 126 141 156 171 184 198 214 230
	32	6 14 20 29 37 46 53 61 70 78 85 93 99 105 114 121 128 135 143 151 159 167 174 181 187 195 202 208 214 221 229 39	16 34 42 56 66 71 75 76 78 85 93 96 101 110 117 129 133 135 141 147 148 158 169 177 193 203 203 214 214 221 222 236	7 15 24 32 40 48 56 64 72 81 89 93 98 105 113 121 129 135 140 144 150 156 164 171 176 183 189 197 207 219 227 234	6 12 20 27 34 41 50 60 66 72 78 83 92 101 108 115 121 127 136 146 157 165 172 179 185 191 198 204 213 221 228 236	4 9 13 18 25 31 37 44 51 59 68 77 87 96 104 111 120 127 134 142 149 157 166 175 184 191 199 208 214 221 229 240
Peppers	2	63 145	55 144	63 145	63 145	63 145
	4	61 113 160 227	46 122 154 195	57 90 154 227	61 113 160 227	56 89 164 227

(continued)

Table 15.5 (continued)

I	nt	ALO	PSO	CSA	RRA	ACOR
	8	34 60 87 113 141 169 198 227	25 69 80 110 140 152 178 219	34 60 87 112 140 168 197 227	32 59 86 120 148 175 198 227	33 59 85 111 140 168 197 227
	16	22 37 52 64 78 92 109 125 141 155 170 184 197 208 222 232	23 52 59 85 115 118 133 157 168 171 188 205 205 221 229 232	7 22 36 50 63 77 90 106 120 134 151 167 183 200 221 231	7 29 45 59 75 89 107 123 138 154 170 185 198 209 223 233	7 21 33 45 59 74 89 105 117 132 148 165 182 200 221 232
	32	7 13 22 30 38 47 56 65 71 79 86 93 103 109 117 123 130 137 146 153 161 168 175 184 193 198 206 213 220 227 233 244	8 15 35 35 41 47 51 54 57 71 73 76 87 90 100 109 112 115 127 134 136 150 166 177 186 189 200 204 204 208 214 226	7 11 17 23 30 40 51 59 67 75 84 92 101 110 120 131 138 145 155 161 169 177 183 190 197 203 209 215 220 227 235 244	6 13 20 26 33 39 44 49 55 63 72 82 89 96 102 114 126 137 147 156 164 170 177 184 190 197 204 211 222 232 238 245	7 13 19 25 32 38 45 52 60 69 76 83 90 98 106 113 120 127 135 142 149 157 166 173 180 188 196 204 214 222 230 239
Butterfly	2	27 213	27 133	27 213	27 213	117 213
	4	27 91 137 213	22 91 117 142	27 91 136 213	27 63 121 213	27 91 137 213
	8	13 27 77 105 131 164 213 233	16 49 58 100 129 146 196 213	27 73 96 119 144 164 213 234	27 73 98 131 170 188 213 234	27 63 87 111 136 164 213 233
	16	9 18 27 61 81 98 117 131 141 152 164 183 213 226 236 247	1 4 11 17 57 62 64 92 107 143 153 189 203 213 226 231	13 27 46 63 76 87 98 109 120 131 148 172 213 221 231 242	4 15 27 47 73 94 113 131 148 163 173 195 213 224 233 242	7 15 26 46 63 76 89 102 116 131 144 163 185 213 226 241

(continued)

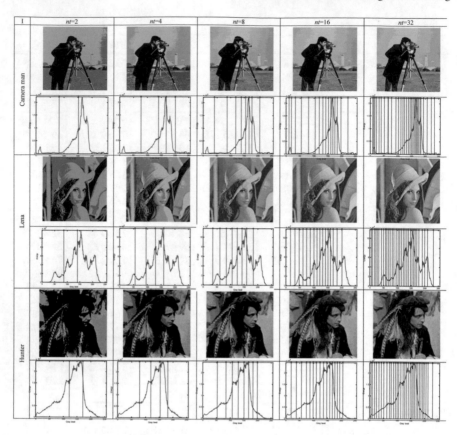

Fig. 15.4 Segmented images using ALO and Otsu's method

Table 15.5 (continued)

I	nt	ALO	PSO	CSA	RRA	ACOR
	32	10 16 22	3 5 20 49	6 9 15 22	6 13 19 28	4 9 15 20
		27 28 36	61 67 67	27 45 54	36 39 42	27 28 37
		48 61 72	68 84 85	61 65 72	57 65 73	47 57 65
		80 93 102	91 92 108	80 90 98	80 88 98	73 82 90
		109 117	115 120	106 112	108 117	97 104 111
		124 131	124 131	121 131	125 134	119 124
		137 146	133 133	140 148	152 164	131 138
		151 156	140 145	156 163	176 182	144 157
		160 173	151 160	174 185	187 202	163 176
		188 198	168 179	200 203	206 212	182 193
		206 212	180 189	213 222	213 221	213 217
		213 220	203 223	225 231	228 235	222 228
		231 237	237 240	236 242	240 245	237 246
		242 249	247	247	250	

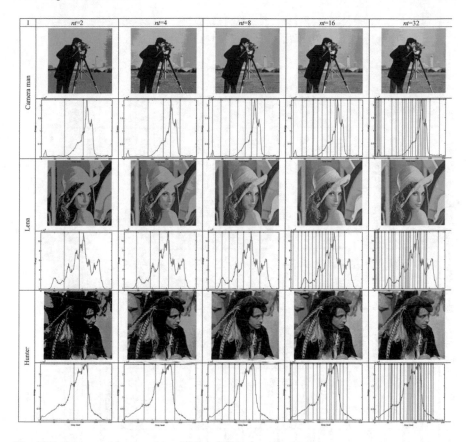

Fig. 15.5 Segmented images using ALO and Kapur's method

Table 15.6 presents the mean and also the standard deviation (σ) of the PSNR values for all algorithms, where a higher value indicates better segmentation quality.

In Table 15.6 it is possible to observe that some algorithms generate high PSNR values. However, this metric alone cannot ensure that the segmentation has been successful. For example, algorithms such the PSO get bad thresholds, but still in some cases get good PSNR values. This happens because the pixels can be incorrectly classified. In addition, Tables 15.7 and 15.8 show the results of the other two metrics that analyze the quality of segmentation.

Now, the segmented images obtained by the ALO methods using Kapur are presented for visual inspection in Fig. 15.5. As in the previous section, only three representative images are shown.

In this subsection Tables 15.6, 15.7 and 15.8 showed evidence of the performance of each of the approaches evaluated. From such results it is possible to conclude that the ALO algorithm is the evaluated method that obtains the best results for the

Table 15.6 Mean and σ of the PSNR values obtained by the ALO, PSO, CSA, RRA and ACOR methods using the Kapur method

I	nt	ALO		PSO		CSA		RRA		ACOR	
		Media	σ	Media	σ	Media	σ	Media	σ	Media	σ
Cameraman	2	14.1078	5.40E−15	**14.1419**	5.40E−02	13.9197	5.40E−15	14.0816	1.70E−01	13.8256	5.41E−15
	4	**20.2737**	1.90E−03	19.9816	7.10E−01	17.238	4.60E−03	20.2462	1.80E−01	19.8685	2.40E−03
	8	**24.5888**	5.50E−01	23.4598	1.60E+00	23.2151	4.40E−01	24.2921	5.90E−01	23.7074	3.59E−01
	16	**29.3025**	6.00E−01	27.1448	1.80E+00	28.7052	8.40E−01	29.1105	8.40E−01	28.7225	7.12E−01
	32	34.674	1.20E+00	31.6333	1.60E+00	34.9571	9.10E−01	34.8934	1.10E+00	**35.3253**	2.98E−01
Lena	2	14.7275	4.70E−05	**14.7292**	5.20E−02	11.9276	9.00E−15	14.7275	3.30E−05	14.433	9.01E−15
	4	**19.1959**	2.60E−01	18.3044	7.90E−01	16.5724	3.60E−03	19.203	1.30E−01	18.7623	3.92E−01
	8	**24.0628**	5.80E−01	22.3626	1.50E+00	22.9017	2.80E−01	24.0068	4.90E−01	23.9697	1.91E−01
	16	**29.9022**	7.90E−01	26.5415	1.40E+00	29.1179	5.60E−01	29.6602	7.20E−01	29.5047	2.90E−01
	32	34.4498	1.30E+00	31.8827	1.40E+00	34.8237	5.40E−01	**35.2422**	8.00E−01	35.129	2.87E−01
Hunter	2	15.2267	2.10E−05	15.2172	2.20E−02	14.8311	1.20E−05	**15.2268**	3.50E−03	14.9222	1.99E−05
	4	**21.126**	1.90E−02	20.797	3.70E−01	18.2206	3.80E−02	21.0829	4.00E−02	20.6954	1.68E−02
	8	**24.6765**	1.90E−01	24.1361	1.30E+00	24.0139	5.50E−01	24.5576	2.40E−01	24.924	5.17E−01
	16	**30.3006**	2.30E−01	27.5317	1.30E+00	29.6918	4.20E−01	30.2806	4.80E−01	30.2479	1.74E−01
	32	35.9282	4.70E−01	32.1929	1.10E+00	35.751	5.90E−01	35.8244	6.40E−01	**36.0226**	2.11E−01
Peppers	2	**16.074**	3.50E−04	16.0244	1.10E−01	11.6776	5.10E−04	16.0684	2.00E−02	15.7381	2.80E−02
	4	19.6974	1.30E+00	19.8763	1.30E−01	18.2128	9.60E−01	18.5166	7.90E−01	**19.9274**	1.11E+00
	8	**24.0135**	1.80E−01	23.0179	1.00E+00	23.4129	8.10E−02	23.8814	6.50E−01	23.5046	2.80E−01
	16	**29.5441**	4.50E−01	26.5244	1.50E+00	28.831	4.00E−01	29.4239	5.00E−01	29.0631	2.43E−01

(continued)

Table 15.6 (continued)

I	nt	ALO		PSO		CSA		RRA		ACOR	
		Media	σ	Media	σ	Media	σ	Media	σ	Media	σ
	32	35.053	5.00E−01	31.5367	1.40E+00	34.8844	4.50E−01	**35.432**	5.00E−01	35.0978	2.36E−01
Butterfly	2	**11.1785**	7.50E−01	11.1757	1.60E+00	10.5827	1.60E+00	10.3281	1.60E+00	10.5759	1.37E+00
	4	**16.8186**	5.90E−01	16.3214	1.50E+00	16.3108	7.30E−02	16.4807	1.10E+00	16.8177	3.52E−02
	8	22.4366	9.70E−01	20.8613	2.00E+00	22.5629	9.80E−01	22.1867	1.00E+00	**22.709**	8.55E−01
	16	27.8741	1.40E+00	25.4166	1.90E+00	**28.5601**	1.10E+00	27.5124	1.10E+00	28.3596	7.06E−01
	32	32.9899	1.20E+00	30.1461	1.80E+00	**34.6828**	1.00E+00	32.7992	1.50E+00	33.5067	6.04E−01

Table 15.7 Mean and σ of SSIM values obtained by ALO, PSO, CSA, RRA and ACOR methods using Kapur's method

I	nt	ALO		PSO		CSA		RRA		ACOR	
		Media	σ	Media	σ	Media	σ	Media	σ	Media	σ
Cameraman	2	**0.7134**	4.51E−16	0.7126	1.37E−03	0.5574	4.50E−16	0.7132	1.40E−03	0.6991	4.51E−16
	4	0.8318	1.96E−04	0.8284	1.99E−02	0.6324	6.80E−04	**0.8326**	2.80E−03	0.8151	1.97E−04
	8	0.883	1.58E−03	0.8687	1.75E−02	0.7343	1.60E−03	**0.8834**	3.80E−03	0.866	1.91E−03
	16	0.9288	2.04E−02	0.9179	2.66E−02	0.8493	2.20E−02	**0.9385**	2.20E−02	0.9223	2.28E−02
	32	0.9773	1.44E−02	0.9614	1.91E−02	0.948	8.40E−03	**0.9788**	1.20E−02	0.9676	1.31E−03
Lena	2	0.6255	2.06E−06	**0.6262**	3.67E−03	0.3927	0.00E+00	0.6255	1.50E−06	0.613	2.25E−16
	4	0.8118	1.63E−02	0.7738	3.78E−02	0.6074	8.70E−05	**0.8146**	7.70E−03	0.7952	1.33E−02
	8	**0.9041**	1.32E−02	0.8758	3.11E−02	0.7992	3.20E−03	0.9035	7.50E−03	0.8897	2.03E−03
	16	**0.9646**	5.86E−03	0.9337	1.73E−02	0.9061	3.90E−03	0.9635	5.20E−03	0.9479	1.72E−03
	32	0.9845	4.86E−03	0.9729	6.87E−03	0.963	1.40E−03	**0.9875**	2.60E−03	0.9694	8.30E−04
Hunter	2	**0.6055**	2.35E−07	0.6017	6.66E−03	0.3445	1.30E−07	0.6046	2.60E−03	0.5934	2.23E−07
	4	**0.8367**	1.71E−03	0.8367	1.36E−02	0.5043	1.90E−03	0.8353	6.90E−03	0.8212	1.78E−03
	8	0.9189	4.69E−03	0.9066	3.85E−02	0.692	7.60E−03	**0.9201**	9.70E−03	0.917	4.43E−03
	16	**0.9733**	1.89E−03	0.9542	1.62E−02	0.8435	4.10E−03	**0.9733**	6.20E−03	0.9611	1.34E−03
	32	**0.993**	1.44E−03	0.9825	5.49E−03	0.944	2.90E−03	0.9912	2.80E−03	0.9758	3.81E−04
Peppers	2	0.7555	5.39E−08	**0.7565**	4.32E−03	0.4955	7.70E−08	0.7555	3.50E−04	0.7403	1.60E−04
	4	0.8334	2.26E−02	**0.8369**	2.48E−02	0.6607	1.70E−02	0.8138	1.60E−02	0.829	1.92E−02
	8	**0.9187**	8.59E−03	0.9026	1.82E−02	0.7686	3.20E−02	0.9168	8.40E−03	0.9035	7.57E−04
	16	**0.9738**	2.46E−03	0.9498	1.16E−02	0.8768	1.90E−03	0.9732	3.00E−03	0.9579	6.66E−04

(continued)

Table 15.7 (continued)

I	nt	ALO		PSO		CSA		RRA		ACOR	
		Media	σ	Media	σ	Media	σ	Media	σ	Media	σ
	32	**0.9935**	8.69E−04	0.9807	5.80E−03	0.95	7.30E−04	0.9928	7.40E−04	0.9747	3.65E−04
Butterfly	2	0.439	3.39E−02	**0.4569**	1.14E−01	0.248	7.30E−02	0.4033	7.70E−02	0.4161	7.17E−02
	4	**0.7909**	2.39E−02	0.7558	7.55E−02	0.5381	3.20E−03	0.7796	5.10E−02	0.7899	6.95E−04
	8	**0.9246**	1.67E−02	0.8865	5.54E−02	0.7633	1.60E−02	0.9189	2.00E−02	0.9191	1.46E−02
	16	**0.9756**	8.76E−03	0.9523	2.47E−02	0.8889	6.50E−03	0.9733	7.90E−03	0.963	3.53E−03
	32	**0.992**	2.80E−03	0.9818	1.02E−02	0.9603	2.10E−03	0.9916	3.80E−03	0.9748	1.27E−03

Table 15.8 Mean and σ of FSIM values obtained by the ALO, PSO, CSA, RRA and ACOR methods using the Kapur's method

I	nt	ALO		PSO		CSA		RRA		ACOR	
		Media	σ	Media	σ	Media	σ	Media	σ	Media	σ
Cameraman	2	0.7229	4.50E−16	0.7208	1.40E−03	0.7063	5.60E−16	**0.7234**	3.50E−03	0.7084	5.63E−16
	4	0.837	2.00E−04	0.8382	2.00E−02	0.7864	8.80E−04	**0.8392**	5.30E−03	0.8202	4.12E−04
	8	**0.9147**	1.60E−03	0.8982	1.70E−02	0.8974	2.70E−03	0.9137	4.50E−03	0.8963	3.42E−03
	16	**0.9612**	2.00E−02	0.9359	2.70E−02	0.953	4.10E−03	0.9583	5.60E−03	0.9411	1.93E−03
	32	0.983	1.40E−02	0.9653	1.90E−02	0.9834	2.40E−03	**0.9836**	3.00E−03	0.9676	9.03E−04
Lena	2	**0.678**	2.10E−06	0.6779	3.70E−03	0.6167	5.60E−16	0.678	5.30E−06	0.6644	5.63E−16
	4	0.758	1.60E−02	**0.7629**	3.80E−02	0.7224	6.00E−05	0.7588	5.20E−03	0.7417	4.85E−03
	8	**0.8857**	1.30E−02	0.848	3.10E−02	0.857	4.50E−03	0.8822	9.70E−03	0.8674	3.26E−03
	16	**0.9511**	5.90E−03	0.9067	1.70E−02	0.9422	5.50E−03	0.9491	5.50E−03	0.9356	2.94E−03
	32	0.9792	4.90E−03	0.9598	6.90E−03	0.9812	1.90E−03	**0.9842**	3.20E−03	0.9673	9.84E−04
Hunter	2	**0.7099**	2.30E−07	0.7075	6.70E−03	0.6677	1.00E−08	0.7094	1.70E−03	0.6957	1.80E−08
	4	**0.8861**	1.70E−03	0.8825	1.40E−02	0.816	8.90E−04	0.8856	3.10E−03	0.869	8.67E−04
	8	**0.9501**	4.70E−03	0.9343	3.80E−02	0.9363	4.50E−03	0.9493	3.90E−03	0.9393	3.92E−03
	16	**0.987**	1.90E−03	0.966	1.60E−02	0.983	2.40E−03	0.9866	3.50E−03	0.9709	7.09E−04
	32	**0.9968**	1.40E−03	0.9874	5.50E−03	0.9962	1.30E−03	0.996	1.50E−03	0.978	1.28E−04
Peppers	2	0.7202	5.40E−08	**0.721**	4.30E−03	0.6238	8.20E−05	0.7204	2.30E−04	0.7058	1.09E−04
	4	0.7965	2.30E−02	**0.8055**	2.50E−02	0.7749	2.20E−02	0.7806	1.60E−02	0.7886	2.61E−02
	8	**0.8959**	8.60E−03	0.8718	1.80E−02	0.8816	7.30E−04	0.8918	8.80E−03	0.8749	5.36E−03
	16	**0.96**	2.50E−03	0.9217	1.20E−02	0.95	3.50E−03	0.9594	3.80E−03	0.9418	1.91E−03

(continued)

Table 15.8 (continued)

I	nt	ALO		PSO		CSA		RRA		ACOR	
		Media	σ	Media	σ	Media	σ	Media	σ	Media	σ
	32	0.9891	8.70E−04	0.9648	5.80E−03	0.9871	1.80E−03	**0.9902**	1.50E−03	0.9723	6.56E−04
Butterfly	2	0.6189	3.40E−02	0.5899	1.10E−01	**0.6353**	1.30E−02	0.5894	3.20E−02	0.5867	3.93E−03
	4	0.7596	2.40E−02	0.7617	7.50E−02	**0.765**	1.60E−03	0.7515	2.40E−02	0.7524	2.08E−04
	8	**0.8907**	1.70E−02	0.8603	5.50E−02	0.8828	1.40E−02	0.8845	1.70E−02	0.8824	1.33E−02
	16	0.9595	8.80E−03	0.925	2.50E−02	**0.9651**	9.20E−03	0.9558	1.10E−02	0.9498	5.48E−03
	32	0.9845	2.80E−03	0.9666	1.00E−02	**0.9922**	4.30E−03	0.9834	8.10E−03	0.97	2.53E−03

thresholding of digital images considering both objective functions defined by Otsu
and Kapur.

15.4.3   Statistical Comparison

This subsection presents a comparative study between the results obtained by the
ALO, PSO, CSA, RRA and ACOR algorithms. Since all these algorithms use random numbers, the results may vary from one evaluation to another. The values of the
objective function of each method are statistically compared with the Wilcoxon non-
parametric significance test, already used previously and which, as in previous cases,
is performed with 35 independent samples. The analysis is carried out considering
a 5% significance on the Otsu values. This test is evaluated on each image considering different numbers of thresholds nt. Table 15.9 reports the p-values produced by
the Wilcoxon tests considering two to two comparisons between: ALO versus PSO,
ALO versus CSA, ALO versus RRA, ALO versus ACOR.

As a null hypothesis it is assumed that there is no difference between the values
generated by the two algorithms in question. The alternative hypothesis considers the
existence of a difference between the values of both methods. Table 15.9 shows that
the p-values for all comparisons are less than 0.05 (5%) which is evidence against the
null hypothesis. This fact indicates that the results obtained by ALO are significantly
different from those of any other evaluated algorithm and did not happen by chance.
In addition, Table 15.10 shows the results of p-values for comparisons between ALO
versus PSO, ALO versus CSA, ALO versus RRA, ALO versus ACOR.

The values shown in Table 15.10 show that the ALO algorithm generates significantly different solutions to the rest of the methods evaluated with the Kapur
objective function. These results together with the quality metrics allow us to affirm
that the ALO algorithm is superior to the algorithms evaluated for the threshold of
digital images.

The results obtained by the proposed algorithms (specifically ALO) are compared
with similar methods as shown in Tables 15.1, 15.2, 15.3 and 15.4 using Otsu and
Tables 15.5, 15.6, 15.7 and 15.8 using Kapur. Such comparisons include graphical
results illustrating the segmentation process performed by the ALO and SCA algorithms on the energy curve on Figs. 15.4 and 15.5. The results in Tables 15.1 and
15.5 indicate that practically all algorithms have the same performance when $nt = 2$.
This situation occurs because the threshold problem contains only two dimensions.
In contrast, when the number of thresholds to be found increases, the algorithms tend
to deliver suboptimal solutions that affect the quality of the segmented images. Considering such facts, the ALO algorithm is able to perform better searches in spaces
of many dimensions, compared to other approaches.

Tables 15.2, 15.3, 15.4, 15.5, 15.6, 15.7 and 15.8 show the quality of the results
of the segmented images using the thresholds obtained by the selected algorithms
and using Otsu and Kapur as the target function. The values of the PSNR, SSIM
and FSIM metrics are increased according to the number of thresholds. Here it is

Table 15.9 p-values of the Wilcoxon test performed on the values of the Otsu target function

I	nt	p-values			
		ALO versus			
		PSO	CSA	RRA	ACOR
Cameraman	2	1.51E−08	4.57E−01	4.24E−02	5.85E-07
	4	1.53E−14	3.81E−01	4.95E−08	5.32E−14
	8	5.71E−13	2.00E−01	4.31E−03	5.70E−13
	16	6.54E−13	2.00E−02	3.07E−02	1.22E−08
	32	1.97E−12	3.78E−01	2.09E−01	4.33E−02
Lena	2	3.49E−06	1.60E−01	5.69E−01	1.50E−06
	4	1.53E−14	6.38E−08	1.19E−06	7.78E−06
	8	1.77E−09	6.59E−01	1.24E−05	2.81E−07
	16	4.56E−12	8.88E−01	1.71E−02	6.14E−01
	32	2.77E−12	2.72E−02	8.15E−04	1.72E−07
Hunter	2	1.49E−08	3.33E−02	1.78E−04	9.85E−06
	4	1.53E−14	9.53E−01	2.85E−09	2.38E−07
	8	5.91E−13	5.06E−04	6.98E−08	3.97E−12
	16	6.54E−13	3.16E−08	1.44E−11	7.42E−01
	32	6.54E−13	9.25E−04	8.21E−02	4.80E−10
Peppers	2	5.56E−09	2.65E−01	3.02E−03	1.43E−06
	4	1.53E−14	4.24E−01	1.34E−01	2.37E−11
	8	2.77E−12	2.40E−01	3.73E−05	1.13E−11
	16	1.19E−12	2.20E−06	8.06E−06	6.81E−01
	32	8.46E−13	7.81E−02	2.96E−01	1.52E−07
Butterfly	2	1.01E−07	6.47E−05	3.98E−01	9.85E−08
	4	1.53E−14	1.39E−07	5.88E−10	1.98E−09
	8	7.13E−08	2.12E−06	5.62E−07	9.44E−04
	16	1.25E−09	4.16E−03	6.90E−03	2.09E−10
	32	1.93E−10	5.79E−06	2.44E−03	7.13E−13

important to mention that the use of the energy curve helps to have more information on the distribution of pixels in the image than when using the histogram. The results achieved by the ALO algorithm are higher in most cases. This behavior occurs because each image has different characteristics that may hinder or simplify the optimization problem. In addition, the randomness of the metaheuristic algorithms generates variation in the results. Finally, based on the NFL theorem [17] and using the results presented in the previous sections, it is possible to conclude that it is difficult to define whether an algorithm can segment any image correctly. This is also supported by the fact that the energy curve is multimodal. For example, if a threshold s in an area of the energy curve that is not appropriate, the segmentation

Table 15.10 *p*-values of the Wilcoxon test performed on the values of the Kapur objective function

I	nt	*p*-values			
		ALO versus			
		PSO	CSA	RRA	ACOR
Cameraman	2	5.60E−09	4.24E−02	4.57E−01	1.77E−01
	4	2.18E−14	4.95E−08	3.81E−01	3.58E−01
	8	7.11E−13	4.31E−03	2.00E−01	2.73E−03
	16	6.55E−13	3.07E−02	2.00E−02	5.81E−01
	32	6.55E−13	2.09E−01	3.78E−01	1.18E−05
Lena	2	6.36E−10	5.69E−01	1.60E−01	1.60E−01
	4	1.83E−12	1.19E−06	6.38E−08	7.76E−04
	8	1.28E−12	1.24E−05	6.59E−01	4.51E−08
	16	6.55E−13	1.71E−02	8.88E−01	3.85E−01
	32	6.55E−13	8.15E−04	2.72E−02	3.27E−12
Hunter	2	2.67E−11	1.78E−04	3.34E−02	9.99E−02
	4	2.13E−13	2.85E−09	9.53E−01	4.73E−01
	8	6.54E−13	6.98E−08	5.06E−04	3.00E−12
	16	6.55E−13	1.44E−11	3.16E−08	9.22E−13
	32	6.55E−13	8.21E−02	9.25E−04	3.56E−05
Peppers	2	2.08E−09	3.02E−03	2.65E−01	2.95E−03
	4	3.53E−11	1.34E−01	4.24E−01	6.57E−06
	8	1.40E−12	3.73E−05	2.40E−01	1.36E−03
	16	6.55E−13	8.06E−06	2.20E−06	1.10E−01
	32	6.55E−13	2.96E−01	7.81E−02	1.24E−02
Butterfly	2	3.85E−11	3.98E−01	6.47E−05	1.72E−01
	4	5.91E−13	5.88E−10	1.39E−07	1.69E−04
	8	6.54E−13	5.62E−07	2.12E−06	7.32E−04
	16	6.55E−13	6.90E−03	4.16E−03	4.36E−05
	32	6.55E−13	2.44E−03	5.79E−06	3.03E−07

will not be the best. Such a situation can only be reflected through specialized metrics such as PSNR, SSIM and FSIM because the target function of both Otsu and Kapur works with limited information.

15.5   Summary

This chapter presents a study on techniques applied to the multi-level thresholding of images incorporating the formulation of the energy curve as a new concept and that allows adding existing contextual information in the vicinity of each pixel of the image, as opposed to the histogram that only contains information on intensity levels. The evaluation is made considering two classic threshold criteria (Otsu and Kapur) where these criteria are applied on the energy curve instead of on the histogram as in its original definitions. The performance of such a formulation is exhaustively evaluated using metaheuristic algorithms where the ALO algorithm shows the best performance.

References

1. Ghosh S, Bruzzone L, Patra S et al (2007) A context-sensitive technique for unsupervised change detection based on hopfield-type neural networks. IEEE Trans Geosci Remote Sens 45:778–789. https://doi.org/10.1109/TGRS.2006.888861
2. Hammouche K, Diaf M, Siarry P (2010) A comparative study of various meta-heuristic techniques applied to the multilevel thresholding problem. Eng Appl Artif Intell 23:676–688. https://doi.org/10.1016/j.engappai.2009.09.011
3. Sezgin M, Sankur B (2004) Survey over image thresholding techniques and quantitative performance evaluation. J Electron Imaging 13:146–166
4. El AM, Ewees AA, Hassanien AE (2017) Whale optimization algorithm and moth-flame optimization for multilevel thresholding image segmentation. Expert Syst Appl 83:242–256
5. Dehshibi MM, Sourizaei M, Fazlali M, et al (2017) A hybrid bio-inspired learning algorithm for image segmentation using multilevel thresholding. Multimed Tools Appl 76. https://doi.org/10.1007/s11042-016-3891-3
6. Hussein WA, Sahran S, Abdullah SNHS (2016) A fast scheme for multilevel thresholding based on a modified bees algorithm. Knowl-Based Syst 101:114–134
7. Chuang L-Y, Yang C-H, Li J-C (2011) Chaotic maps based on binary particle swarm optimization for feature selection. Appl Soft Comput 11:239–248. https://doi.org/10.1016/j.asoc.2009.11.014
8. Suresh S, Lal S (2017) Multilevel thresholding based on chaotic darwinian particle swarm optimization for segmentation of satellite images. Appl Soft Comput 55:503–522. https://doi.org/10.1016/j.asoc.2017.02.005
9. Pare S, Kumar A, Bajaj V, Singh GK (2016) A multilevel color image segmentation technique based on cuckoo search algorithm and energy curve. Appl Soft Comput 47:76–102. https://doi.org/10.1016/J.ASOC.2016.05.040
10. Pare S, Bhandari AK, Kumar A, Singh GK (2017) An optimal color image multilevel thresholding technique using grey-level co-occurrence matrix. Expert Syst Appl 87:335–362. https://doi.org/10.1016/J.ESWA.2017.06.021
11. Otsu N (1979) A threshold selection method from gray-level histograms. IEEE Trans Syst Man Cybern 9:62–66. https://doi.org/10.1109/TSMC.1979.4310076
12. Kapur JN, Sahoo PK, Wong AKC (1985) A new method for gray-level picture thresholding using the entropy of the histogram. Comput Vis Graph Image Process 29:273–285. https://doi.org/10.1016/0734-189X(85)90125-2
13. Kennedy J, Eberhart RC (1995) Particle swarm optimization. In: Proceedings of IEEE international conference on neural networks, vol 4, pp 1942–1948. https://doi.org/10.1109/ICNN.1995.488968
14. Askarzadeh A (2016) A novel metaheuristic method for solving constrained engineering optimization problems: crow search algorithm. Comput Struct 169:1–12. https://doi.org/10.1016/j.compstruc.2016.03.001
15. Merrikh-Bayat F (2015) The runner-root algorithm: a metaheuristic for solving unimodal and multimodal optimization problems inspired by runners and roots of plants in nature. Appl Soft Comput 33:292–303. https://doi.org/10.1016/J.ASOC.2015.04.048

16. Socha K, Dorigo M (2008) Ant colony optimization for continuous domains. Eur J Oper Res 185:1155–1173. https://doi.org/10.1016/j.ejor.2006.06.046
17. Wolpert DH, Macready WG (1997) No free lunch theorems for optimization. IEEE Trans Evol Comput 1:67–82. https://doi.org/10.1109/4235.585893
18. Ghamisi P, Couceiro MS, Benediktsson JA, Ferreira NMF (2012) An efficient method for segmentation of images based on fractional calculus and natural selection. Expert Syst Appl 39:12407–12417. https://doi.org/10.1016/j.eswa.2012.04.078
19. Akay BB (2013) A study on particle swarm optimization and artificial bee colony algorithms for multilevel thresholding. Appl Soft Comput 13:3066–3091. https://doi.org/10.1016/j.asoc. 2012.03.072
20. Il-Seok O, Lee J-S, Moon B-R (2004) Hybrid genetic algorithms for feature selection. IEEE Trans Pattern Anal Mach Intell 26:1424–1437. https://doi.org/10.1109/TPAMI.2004.105
21. Agrawal S, Panda R, Bhuyan S, Panigrahi BK (2013) Tsallis entropy based optimal multilevel thresholding using cuckoo search algorithm. Swarm Evol Comput 11:16–30. https://doi.org/ 10.1016/j.swevo.2013.02.001
22. Horng M-H, Liou R-J (2011) Multilevel minimum cross entropy threshold selection based on the firefly algorithm. Expert Syst Appl 38:14805–14811. https://doi.org/10.1016/j.eswa.2011. 05.069
23. Wang Z, Bovik ACAC, Sheikh HRHR, Simoncelli EPEP (2004) Image quality assessment: from error visibility to structural similarity. IEEE Trans Image Process 13:600–612. https:// doi.org/10.1109/TIP.2003.819861
24. Zhang L, Zhang L, XuanqinMou DZ (2011) FSIM: a feature similarity index for image. IEEE Trans Image Process 20:2378–2386

Printed in the United States
By Bookmasters